ADD HUMAN

HUMAN FACTORS & ERGONOMICS FOR ALL OF US

THOMAS WAYNE KING

CONTENTS

Welcome to ADD HUMAN	ix
Preface And Background	xiii
1. ADD HUMAN	1
HUMAN FACTORS *ARE*…	4
HUMAN FACTORS *ARE NOT*…	7
HUMAN FACTORS AND ERGONOMICS IN DAILY LIFE	9
QUICK HUMAN FACTORS QUIZ	10
REVIEW QUESTIONS	16
2. YOU AND ME AND HUMAN FACTORS	17
CASE EXAMPLES	19
Transparency versus Opacity	19
Mappings	20
Affordances	21
Feedback from Switches	22
A USEFUL FRAMEWORK: BAKER'S BASIC ERGONOMIC EQUATION	23
Motivation.	23
Physical Effort	25
Cognitive Effort	25
Sensory Effort	26
Linguistic Effort	26
Time Load	27
CASE EXAMPLES	28
Motivation to Pursue and Complete a Task	28
Physical Effort	28
Cognitive Load	29
Linguistic Load	29
Time	30
SUMMARY	31
REVIEW QUESTIONS	32

3. DESIGNING FOR PEOPLE	33
BENNY: An instructive case example.	38
ENTER HUMAN FACTORS	39
TECH TRANSPARENCY…AND MORE	43
ASSISTIVE TECHNOLOGY & ERGONOMICS FOR ALL OF US	47
BRIEF HISTORY OF ASSISTIVE TECHNOLOGY	48
COMPONENT AREAS OF ASSISTIVE TECHNOLOGY	52
1. *Augmentative and Alternative Communication (AAC)*	53
2. *Adapted Computer Access*	54
3. *Assistive Listening and Seeing Devices*	56
4. *Environmental Control*	56
5. *Adapted Play and Recreation*	57
6. *Seating and Positioning*	58
7. *Mobility and Powered Mobility*	59
8. *Prosthetics*	60
9. *Rehabilitation Robotics*	61
10. *Integration of Assistive Technology into Home, School, and Community*	61
PRIMARY USERS OF ASSISTIVE TECHNOLOGY	62
TABLE 3-1. SPECIAL NEEDS & REPRESENTATIVE USES OF ASSISTIVE TECHNOLOGY	65
THOSE WHO WORK WITH ASSISTIVE TECHNOLOGY	66
WHO ARE USERS OF ASSISTIVE TECHNOLOGY?	72
CASE EXAMPLES	73
Lynn	73
Ron	74
SUMMARY	76
REVIEW QUESTIONS	77
4. HUMAN FACTORS GOALS	79
HUMAN FACTORS AND INTENTS OF ASSISTIVE TECHNOLOGY	83
APPLICATIONS OF BAKER'S BASIC ERGONOMIC EQUATION: BBEE	85
Enhancing User Motivation	86
Reducing Physical Effort	90

Reducing Cognitive Effort	93
Reducing Linguistic Effort	94
Reducing Time Load	96
SUMMARY	104
REVIEW QUESTIONS	105

5. REFINING HUMAN FACTORS IDEAS — 107
   SPECIFIC HUMAN FACTORS IN AT — 108

Transparency and Visibility	110
Cosmesis	112
Mappings	116
Affordances	119
Learned or Taught Helplessness	122
Feedback	124
Knowledge of Technology Use Within the User or Within the Device	126
Constraints of Tech Use	129
Forcing or Fail-Safe Functions	132
Preventing Mistakes and Errors	133
SUMMARY	135
REVIEW QUESTIONS	135

6. SWITCHES AND CONTROLS: YOUR PASS KEYS TO TECHNOLOGY — 137
   SWITCHES AND CONTROLS: DEFINITIONS AND EXAMPLES — 138
   SPECIFIC HUMAN FACTORS REGARDING SWITCHES AND CONTROLS — 143

Sensitivity	143
Selectivity	148
Resolution	151
Contrast	155
Feedback	157
Latching and Variability	159
Positioning and Mounting	161
Composition and Construction	163
Cosmesis	169
Sensory Defensiveness of the User	173
A BRIEF INTERNATIONAL PERSPECTIVE ON SWITCHES AND CONTROLS	176
ASSISTIVE TECHNOLOGY AND FITTS' LAW	178
Physical Effort	180

Cognitive and Linguistic Effort Factors	187
Time	191
SUMMARY	194
REVIEW QUESTIONS	195
7. SCREENS: YOUR GATEWAYS TO TECHNOLOGY	197
HUMAN FACTORS RELATED TO SCREENS	208
Physical Size and Bulk of the Screen	209
Screen Quality of Visual Display Information	213
Contrast, Brightness, Resolution, Flicker and Focus	214
Screen Glare and Reflectivity	219
Proxemics, Seating and Positioning	236
SUMMARY	241
REVIEW QUESTIONS	241
8. TECHNOLOGY LEVELS, LITERACY AND LIFE SPAN	243
LEVELS OF ASSISTIVE TECHNOLOGY	244
No-Tech AT Systems	244
Low-Tech AT Systems	245
High-Tech AT Systems	246
Scenario 1	247
Scenario 2	248
Scenario 3	249
TOWARD TECHNOLOGICAL LITERACY	261
ASSISTIVE TECHNOLOGY DIAGNOSIS AND INTERVENTION	269
Assessment	269
Evaluation	270
Diagnosis	270
Intervention	271
TECHNOLOGY ISSUES ACROSS OUR LIFE SPAN	276
Early Childhood	277
Childhood and Early Teen Years	279
Young Adults and Middle-Aged Adults	281
Older Adults	283
SUMMARY	289
REVIEW QUESTIONS	289

9. WHEN ASSISTIVE TECHNOLOGY FAILS	292
FACTORS RELATED TO THE PEOPLE WHO SURROUND THE USER	296
Evaluation and Communication of Diagnostic Information	297
Support and Training	299
Practicalities of Funding	301
Service Delivery and Scheduling	303
Cultural Factors	306
FACTORS RELATED TO THE AT USER	309
Matching Device Features to User (Not User to Device)	309
Cultural Factors	311
Age	312
Transitions among Settings and Devices	313
Literacy Potential	313
Gender	315
FACTORS RELATED TO THE AT DEVICE ITSELF	316
Inattention to Essential Human Factors in Design and Acquisition	316
Mechanical and Electrical Safety and Functioning	320
Weight and Size	322
Power Supply	322
Durability and Repairability	324
FINAL COMMENTS: LOOKING OVER THE HORIZON	325
REVIEW QUESTIONS	330
RESOURCES	333
BIBLIOGRAPHY & LISTING OF ONLINE SOURCES	337
TERMS AND VOCABULARY	369
YOUR HUMAN FACTORS TOOLKIT	379
About the Author	383
Some Other Works by Thomas Wayne King	385
MY TRIBUTE TO MALCOLM "MAL" HANCOCK	387

WELCOME TO ADD HUMAN

Humans use tools. When we succeed with a new device, we are encouraged and feel empowered. When we make errors with a tool or technology, or give up or get hurt because we cannot figure out a device, we blame ourselves. We feel stupid and defeated. Human factors and ergonomics research and practice in engineering and design explore improvement of all matchups of technologies plus human…

ADD HUMAN ~ *human factors and ergonomics for all of us* ~

Practical, applied concepts in human factors and ergonomics across consumer and assistive devices, tools, and technologies.

Thomas Wayne King, Ed.D. (CCC-SLP/L, retired)
Professor Emeritus of Communication Sciences and Disorders, and retired clinical speech-language pathologist. University of Wisconsin- Eau Claire, WI USA

Copyright 2022 Thomas Wayne King. All Rights Reserved.

ISBN 9798496622639

Previously published in 1999 by Allyn & Bacon, then Pearson, Inc. as *ASSISTIVE TECHNOLOGY: Essential Human Factors*. ISBN 0-205-27326-2. *ADD HUMAN* is the updated, enhanced version.

No part of the material protected by this copyright notice may be reproduced or utilized in any form or by any means, electronic or mechanical, including photocopying, recording, or by any information storage and retrieval system, without written permission from the copyright owner.

Photographs by Richard A. Mickelson and Thomas Wayne King.

Line drawings by Gene J. Leisz.

Interior formatting by Hannah Linder Designs

Many of the designations used by manufacturers and others to distinguish their products or services are claimed as trademarks. Where those designations appear in this book, and the author was aware of a trademark claim, the designations have been printed in initial caps or all caps.

Printed in the United States of America.

Contact: Thomas Wayne King
Sunny Cove Publishing PO Box 98,
Solon Springs, WI USA 54873-0098 twkingwrites@gmail.com

Dedicated to Debbi, Seth, and Adam for your support and patience over our many years. Also dedicated to my parents, Victor and Madeline, and to my sister Karen for all that you have taught me about humans using, living with, and adapting devices, tools, and technologies to deal with daunting challenges and life circumstances. Dedicated also to the excellent librarians and libraries everywhere who have encouraged and assisted in making my books happen.

PREFACE AND BACKGROUND

Unexpected ironies of real life can be formative. As I worked on this revised manuscript, I had my own life-changing gravity storm on October 26, 2021, affecting how I use devices in my daily life. In a classic cartoon slip-on-the-unseen-banana-peel flop and dull thud, (on an early-autumn, sloping deck-ramp frost), I dropped hard, flying feet-up with back down, onto frozen boards and ground. With ribs cracked, muscles torn, right arm bruised into swollen purple-yellow-green disfunction, I coped, grateful damage was not worse. What I learned during my months of recovery about reliance on tools, devices, and technologies, simple to complex, clothing to computers to cars, has amplified my enthusiasm for this revision. Each of us adapts to our own tech world as best we can, or are required to. These past months of relearning have been sobering, and are major influences on what enhancements you will read here.

In brief: *We are all temporarily abled*. Things can and do change... slowly, or sometimes in a flash. My brutal reminder now fuels these chapters, with hope that you will engage in ways that reinforce how dependent we all are on things working as expected in our bodies, minds, and tech of daily lives. Access to and successful, safe use under diverse conditions of every item we create, buy, or utilize must be our goals. Human factors and ergonomics considerations apply to all of us, at all ages, in all health and ability conditions. As a clinician, scientist, and educator, but surely not as an engineer, I write now as a more experienced co-participant rather than a professional expert. My recent jolting lessons, offered free by Earth at 1 Gravity, have enlightened and personalized my insights into human factors and ergonomics challenges we all encounter.

In 1999, my then-new book, *Assistive Technology: Essential Human Factors*, was released initially by Allyn & Bacon, Boston, and eventually by Pearson, Inc. This is my 2022 revision of that book, an enhanced version roiling in my thoughts for years. As I worked on this new title, I found updating necessary. But I also found much of what I had written and covered in the first book was still true. Humans are humans throughout the ages. Despite new tools, devices, and tech that we invent and use, human factors considerations hold much the same for all of us over time.

Throughout my experiences and professional practice with persons living with diverse challenges, it remained evident that something in our thinking and training was missing. I had suspected for a long time that there was more to understanding technology needs and preferences of people living with normal-range skills to severe disabilities than what I learned in school or textbooks. Something important like human factors existed, it seemed, but I did not have words for it. No courses, books or instructors mentioned them.

My realizations and new learning began decades ago with information from the Trace Research and Development Center at UW-Madison and from the classic Sanders and McCormick text on human factors. These resources resonated with my personal, family, and professional experiences, and with discussions for years with consumers, students, clients, families, and clinicians. We were all missing something when we dealt with tools, devices, and the other things of tech and tools. Human factors was it. I had found the field that made all these concerns make sense. The human factors view of technology explained so much of what was missing.

That was my motivation in my original writing. It is again now in revising this approachable book on applied human factors. Learners and consumers from a variety of fields need to learn about human factors. So do working education and business professionals, clinicians of many kinds and general consumers, along with assistive technology (AT) users and their families. We all need to *know* and then to *notice* more about the role of human factors in our lives, especially in the lives of our patients, clients, students, and other users who may not advocate for themselves. The available research and published empirical data in human factors applied specifically to AT are emerging and complex, yet book strives to be people intensive and readable. Much of its content comes from my knowing and working with students, patients, clients, and their families. It also comes from decades of contacts and intense discussions with colleagues and friends across many fields of experience and investigation, wise practitioners central to creation of this textbook.

It became apparent to me as a working clinician, university professor, and supervisor of augmentative and alternative communication clinical practice that these tech issues had to be covered in a more formal yet understandable form for people in a diverse variety of backgrounds. Human factors were not talked about in the speech-language pathology, special education, or other curricula of which I was aware, yet the success of AT and all other device use could rise or fall based on our own and our students' knowledge or ignorance of them. I began including the basic content of this book in my courses in the early 1990s and amplified and revised the information continually since. This revision and updating of my original book comes to help fill the still-extant human factors knowledge gap for many of us who know and work with technology and persons of all varieties of abilities. I have tried to offer a useful, practical style, plus an extensive bibliography and summary of contact information for organizations that serve as resources for and links to considerable further consumer learning in human factors and assistive technology.

This revised textbook is my attempt to fill a void in human factors awareness for new learners including diverse university students, working teachers, therapists, and clinicians of all types, as well as for general consumers. It is also and especially for AT users, their families, and supporters. Dr. Jennifer Adler, CUNY cognitive scientist, has given me considerable inspiration, encouragement, and support, and warrants my sincere gratitude. This new approach continues my efforts to help us all to *know* and *notice* more of our built and technological environments. It is for all those who are inquisitive and need to use their knowledge now…not just for specialists. Such innovation can be fraught with perils, and I alone am responsible for errors, oversights, and shortcomings in this book.

Many people were important in the development of the original book and deserve credit for its becoming reality in 1998. For guidance, I especially thank my editor at Allyn and Bacon/Pearson, Inc., Boston, Steve Dragin, and reviewers, Pat Ourand, Rehab Networking, and Diane Chapman Bahr, Loyola College, Maryland.

Dr. Len Gibbs was outstanding in encouragement and motivation offered. Dr. Sylvia Steiner helped immeasurably with encouragement and support for my requests for reassigned time support and sabbatical leave to complete my writing projects. Edie Mateja typed the tables and offered help in manuscript preparation. Florence Clickner made bibliography typing and organization a true art. Gene Leisz did the line drawings with his outstanding flare for excellence. Rick Mickelson was photographer ready at to catch the ideal shot. Ronald Satz and Marjorie Smelstor gave administrative and moral support to the first book project, 1996-1998.

For inspiration, ideas, and stimulation of my professional thinking about human factors and AT, I am also grateful to all the following dedicated, capable people who have lifted my work in so many ways. They are listed here in alphabetical order: Jennifer Adler, Bruce Baker, Brandee Henson Christenson, Tracy Eggert, Pam Elder, Marlene Foster, Zoe Foster, David Franks, Anne Gallatin, Ruth Gullerude, Becky Halverson, Bob Heath, Adam King, Debbi King, Seth King, Darci Lynds, Jim Lynds, Mary Lynds, Julie Maro, Vince Maro, Al Noll, Mark Novak, Pat Ourand, Penny Reed, Barry Romich, Valerie Schauer, Howard Shane, Jen Sonnentag, Robert Sorkin, Trace Center folks, Alex Truesdell, Craig Wadsworth, Jeff White, Ryan White, and Mary Wirkus. Your excellent work and conscientiousness made the first book happen and enabled this update. My appreciation is also extended to all my other colleagues at the University of Wisconsin-Eau Claire and within my writers' community, plus the AT professionals who have make this book and revision into realities.

Special mention is due to all my colleagues in the Department of Communication Disorders at the University of Wisconsin-Eau Claire, 1988-2006, who supported my requests for reassigned time and sabbatical leave to write. Thanks to all, wherever you are now. I could not have done the first book and this revision, ADD HUMAN, without your support, inspiration, contribution of ideas, and tolerance of my frantic days in textbook preparation.

Exceptional recognition and appreciation always go to my amazing wife, best friend, and partner in all I do, Debra Raye King. Debbi, your enthusiasm and encouragement see me through long days on this and all my projects when I think they may never happen. Thanks for all. Our sons, Adam and Seth King offer inspiration, debate, encouragement, technical support, and critique of ideas and formats that have also helped these books become real. Thanks for teaching me so much, guys. I sure appreciate you.

Finally, for my parents, Victor and Madeline, and for my sister, Karen, I offer eternal, heartfelt gratitude for all you have taught me, and for showing me the courage and fortitude it takes to face lives that have been harder, more challenging, and lonelier than most of us can imagine. You have been my source of learning, strength, and inspiration throughout my career. Your impact on my life has been profound. I appreciate all you did for me, hoping this book and my work in human factors with enabling technologies for all can give back some of what you gave me in understanding lives of severe challenges, illness, and disabilities. I am revising this book using much of what you taught me over many years.

- TWK October 31, 2021 -

1

ADD HUMAN

> Five-year-old: *"I can't open this vitamins bottle. It doesn't like me."*
> Mom: *"It has a safety top so it won't open up for little kids."*
> Five-year-old: *"How does it know me?"*

Humans use tools and devices. When we succeed with a new item, we are encouraged and feel empowered. When we make errors with a technology or give up because we get hurt or cannot figure out a device, we blame ourselves. We feel stupid and defeated. Human factors and ergonomics research and practice in engineering and design explore, describe, and seek to improve all matchups of human plus tech. Tools and technologies, simple to complex, are valuable only if we experience effective, safe use without undue errors or discouragement. From toothbrushes to laptops to space craft, celebrated devices are of limited value if people do not successfully interact with them. An awareness and working knowledge of human factors in daily life is important for everyone, regardless of challenges, age, or disability. Technologies of all types may appear useful, but their true worth emerges only as we add the human user.

Human factors concepts and applications are numerous and varied across our lifespans. Consider a few plausible scenarios to illustrate. You may have experienced something like these, or perhaps can imagine them for a few moments to set the background of this book.

Set out several reems of paper on your desk, along with a dozen pens and your laptop computer. Step back and wait a year. Will it become a novel or memoir? No. You must add human. You need someone with the skills, knowledge and motivaton to write and persist with it. Hopefully, you. Get going.

Or place a dozen skeins of your favorite yarn and a pair of sturdy knitting needles on your work counter. Again, step back and wait a month or two. Will it become a new sweater? Not likely. You must add human: a person who has the skills and knowledge, plus the drive, eyes, hands, fingers, and patience to accomplish the intended results.

Or that pile of new boards in lingering your shop, plus your circular saw and box of wood screws...remember? They become new bookshelves only if the builder has time, plus the cognitive, physical, linguistic, and sensory skills to match with the tools to accomplish the project.

Human factors and ergonomics considerations enable safe, successful technology use in our work. Obvious points, sure, but ones we often overlook. Across time, cultures and ethnicities, we humans are regular users of technologies, tools, and devices to accomplish tasks, big and small. And our tools work best when they are engineered and designed to include and adapt to our human capabilities and limitations, however varied.

When devices are attuned to the physical, cognitve, sensory, linguistic and time requirements of the people who will rely on them, then technologies that are efficient and safe will work well, get used, and become part of our acumen. Tools and technologies that do not match with users' capabilities, and that violate human factors considerations we all share, will be avoided and abandoned, even becoming dangerous to users, onlookers, and their surrounding enviroments.

Human factors, the detailed, specific variables surrounding how people operate and utilize devices and tools, are of central importance to success with technologies of all types. Also referred to as *ergonomics*, human factors are often unrecognized, unaddressed aspects of general consumer products and devices, as well as more specialized occupational and assistive technologies (AT). As consumer and enabling technologies developed in scope and effectiveness over the past many decades, it has become evident that focusing on the technology alone is not enough for efficacious results (Anderson, 2021; Proctor and Zandt, 2008; Spacey, 2021).

Almost always, there are other influences on if, how, and how well a tool user may join efforts with a given device to become a true *system*—a human using a device for a purpose. These species-specific influences and aspects of how human beings accept, learn, and effectively use or fail with tools and devices are human factors. Developing basic understandings of and appreciation for essential human factors in common technologies we encounter across varied fields of human endeavor are the focus of this textbook: a good, practical start.

To begin, the term *human factors* seems preferred in North America, with its synonymous term *ergonomics* used more in Europe, but that is not rigid. Related nomenclature for this field includes *human factors in design and engineering, human engineering,* and *engineering psychology,* among others (American Psychological Association, 2014; Kantowitz & Sorkin, 1983; Sanders & McCormick, 1993). By whatever name, human factors research and practice focus on how human beings interact with devices, tools, and technologies in all aspects of play, education, work, and daily living (Anderson, 2021; King, Schomisch, & King, 1996; Masterson, 2021).

Interest in human factors has grown as more technologies of all types, simple to complex, permeate our lives. Those of us in various engineering, consumer, education, and clinical fields recognize that the degree of success that our clients, students, and patients will or will not have with technologies often depends on variables most professionals had not previously considered or been trained in. There is more to our human use of technology than just its alure and our desire to use it. Tech success is also about specific operational details of how we humans will or will not interact with devices and tools we try to utilize.

100 words about the *HUMAN FACTORS AND ERGONOMICS SOCIETY (HFES)*

Human factors and ergonomics are diverse, growing fields of research and practice including professionals from a variety of disciplines. In the United States, the Human Factors and Ergonomics Society, based in Washington, DC, serves as the professional coordinating and information-dissemination organization for human factors and ergonomics (HFE). The society publishes a journal as well as an annual yearbook and directory of engineers, clinicians, and others who teach, research, and practice in the fields of human factors and ergonomics. See www.hfes.org

HUMAN FACTORS *ARE...*

A foundational understanding of human factors is essential as we all rely increasingly on technologies for work, recreation, and daily life, and to heal and make whole. We must recognize what human factors are and what they are not. Anderson (2021), Sanders and McCormick (1993), as well as Kantowitz and Sorkin (1983), among others, have written on this topic, detailing at length the potential role of human factors in design and interaction for general consumers. Based on the descriptions of these authors as well as those of Norman (1988), Stoll (1995), and others, we will look in much more detail at what human factors are and are not. We will attempt to extend and expand on some of the key points of the aforementioned authorities to apply them to the more specialized realms of assistive technologies.

1. Human factors in a general sense is concerned with the many aspects of how all humans interact with devices and technologies. Also, human factors in specialized and assistive technologies (AT) must be concerned with how people who have special needs, limitations, and/or

disabilities interact with common and specialized devices and tools that may support, supplement, or replace some process or ability that has been lost or impaired by illness, injury, or aging. We must be concerned not only with how users of a range of abilities interact with the device, but also with how the families or other close care providers react to the use of tools and devices in their setting. Because persons with limitations who are AT users must closely and frequently interact with or depend on others around them for more of their daily care and other aspects of their lives than general consumers do, the interaction of the AT user alone with the technology introduced into his or her life does not tell us the whole story. The larger impact on those around the user must also be considered because they are key players in implementation of any assistive technology in the user's life. Across all component areas of AT, human factors must be concerned with how the potential user, as well as his or her family, personal care providers, education and therapy aides, teachers, employers and clinicians accept, interact, and live daily with assistive devices and technological systems.
2. Human factors is also concerned with how the design of things influences people, and especially those with exceptional needs for this book. Although it is expanding with our aging population, the market base for AT product development has been smaller than that for general consumers, so not as much attention has been paid to researching the impact of design on user acceptance and use in AT. This is improving. How devices are designed and what materials they are made from can be important factors for the specialized AT consumer, just as these factors are essential to consumers of general technologies.
3. Human factors is concerned with finding out the special needs, capabilities, and limitations of all potential users, and then matching devices and controls to each individual user as best possible: a daunting design task. Heterogeneity and individualization are primary considerations in dealing with

persons who have special needs, yet most often mass-produced technologies must be designed for the widely diverse characteristics of mass-market users, while still empowering unique individuals and their specific needs. Device adaptation and modification for configuration for an individual user with highly specific needs become important considerations in AT—more so than for general consumer technologies. Flexibility and adaptability to a wide range of user characteristics are important.

4. Human factors is focused on increasing a user's effectiveness, efficiency, and convenience with a device or tool. These three elements are always primary factors in the design of products for human beings. *Effectiveness* is the ability of the device to accomplish the task for which it is intended in conjunction with the user. *Efficiency* measures accurate and successful uses versus error uses, plus rate of use. *Convenience* is defined as user ease in application of the device—how much or little effort and discomfort across several domains (physical, cognitive, sensory, linguistic, etc.) and settings is required of the user.

5. Human factors has particular focus on maximizing the user's comfort with a device or system, the satisfaction with the type of work or product of effort that is created by the system, and the overall performance of the consumer or AT device or system. Does it accomplish what they believe was intended for the device or system when they acquired it?

6. Human factors considerations have special interest in reducing the user's exertion, stress, and fear of use. We all have a bit of "technophobia" when it comes to use of new tools and devices, especially complex, high-technology devices. These fears and stresses, particularly when they relate to devices that may be difficult to set up and/or require a great deal of exertion to use, can be deleterious to the AT user. Persons relying on AT typically already have some type of limitation or disability. The expectation to become skilled in using additional tools, devices, or technology can be highly stressful to users because it simply

adds to the frustrations of the handicapping condition already present.
7. Human factors also focuses on complexities of reducing danger to the user and persons around them from the device, as well as the user's possible failure during use, and subsequent rejection of the system for future use—even when it seems that the potential system has considerable merit for an AT user. It is "not just about people 'taking more care'" per Anderson (2021).

HUMAN FACTORS *ARE NOT...*

In addition to considering what human factors are, it is also instructive to discuss what they are *not* (Sanders & McCormick, 1993). Human factors, when understood at just a surface level, can seem so evident that the distinction between what human factors includes and what this area of expertise does not include must be more tightly drawn. As adapted from Anderson (2021), Kantowitz and Sorkin (1983), Norman (1988), and particularly Sanders and McCormick (1993), the following points cover some important areas of what human factors *DO NOT* include.

1. Human factors in usual consumer and assistive technologies are not necessarily obvious, transparent, or visible to potential users; and perhaps especially not to AT designers, engineers, family, teachers, and clinicians who do not have the same special needs or limitations as the clientele with whom they work. The design, configuration, and implementation of AT and its surrounding human factors must be done from the *user's* perspective, not from the viewpoint of some expert but external inventor, viewer, or user. (This is hard to do, by the way.) As Kantowitz and Sorkin (1983) have emphatically pointed out, the primary rule of human factors is to "Honor thy user!" This is particularly true when consumers and users have significant special needs, limitations, disabilities, or challenges.
2. Human factors in AT are not just common sense. The

"common sense" that we tend to develop over years of life experience regarding tool and device use is based on our idiosyncratic and usually normal-range sensory and mobility status, as well as our interactions with devices based on our typically normal-range manipulation skills. When users who have special needs and skills become the focus of our human factors considerations, these elements must evolve to include the unique, yet varied characteristics of this population of actual users.

3. Human factors does not mean using just one person (oneself) as a model for design, selection, configuration, or implementation. This is a trap that we clinicians and educators can easily fall into. Our tendency to choose to work with devices that are designed and configured in ways pleasing to us may lead us away from looking at the special human factors considerations that should be entertained for our intended consumers, as well as clients and students who have unique challenges and limitations.

4. Human factors in AT are not always identical to factors considered in the development and use of general consumer products. Again, persons with special sensory, motoric, or other needs may not be able to interact with devices in the ways we take for granted with the general consumer population—on whom designs of most products are standardized.

---

100 words about *HUMAN FACTORS DESIGNS*

Consumers enjoy devices and tools that work well. Yet some devices are not user friendly, do not work, or are dangerous. Since 1996, Michael J. Darnel has depicted frustrating tools and technologies, simple to complex, that violate human factors considerations. His online scrapbook, *Bad Human Factors Designs*, is a museum of everyday devices, tools and technologies that frustrate us and can even cause harm. Spend a few minutes

browsing. You will soon garner what human factors and ergonomics are about as you observe familiar items, and think of other technologies to include in Darnel's clever, disturbing, fascinating list. See www.baddesigns.com

---

HUMAN FACTORS AND ERGONOMICS IN DAILY LIFE

If you have unknowingly set your alarm for the wrong wake-up time, even though you were convinced the night before that you had set it correctly, or if you have ever found your typing of *wordorsentencespacing onaprintedform* to look something like this so that the text was hard to decipher, or if you have chilled or nearly scalded yourself in a motel room shower, then you have indeed come in contact with things designed with and without human factors in mind. Every time we use a device, tool, or technology, we, at some level of awareness, evaluate it in terms of how "hard" it was to use. This is a common occurrence for all of us, something that we probably deal with several times each day as normal-range ability users of tens, maybe even hundreds of devices, tools, and technologies.

Now, imagine that you have various difference or limitations in how you move, see, hear, or think, and you can begin to understand the importance of incorporating human factors principles in design. Our success—or lack of it—in using technology is often rooted in the human factors aspects of the design of the technology and how we relate to it—more so than in the product or the "gee whiz" potential of the specific devices we are trying to use (King, 1999; Thaler and Sunstein, 2008).

---

100 words about *APPAREL AND HUMAN FACTORS*

Apparel may be our most-encountered technology. Clothing, footwear, and headwear are essential for daily life, combining multiple human factors. Work overalls, aprons, construction boots and safety helmets

are important, but so are our usual shirts, skirts, pants, dresses, vests, socks, shoes, and jackets. We rely on them for warmth, protection, adornment, inclusion, and role identification in personal, social, and occupational settings. Adapted garments, with inclusive cosmesis, styling, fit, closures and fasteners, are essential for people of different sizes, ages, genders, and abilities, enhancing our active participation in life. See **Apparel at Disabled World.** www.disabled-world.com

QUICK HUMAN FACTORS QUIZ

We interact with and rely on numerous devices daily, often unaware of the specific human factors that may be present—or lacking—in their design and operation. We may become aware that something is annoying or hard to use, or that it is easy to use, but are not consciously sure just why that is.

Based on concepts explored by Helander (2006), Smith (1981) as described in Kantowitz and Sorkin (1983), and King, (1999), among others, here is a brief, fun quiz to highlight our intuitive knowledge of, reliance on, and response to certain essential human factors in some general consumer technologies as well as in some aspects of AT. The fundamental human factors considerations depicted here, plus several others, will be described in more detail in the latter portions of this chapter. If you want to, use your pencil to write or sketch responses to each of these based on your knowledge and experience before we go into further discussion of specific human factors.

1. You are sitting in a powered wheelchair looking at this joystick. Which direction do you want the chair to move when you push/pull the control toward each of the arrows? Why? (Mappings, Feedback…)

ADD HUMAN | 11

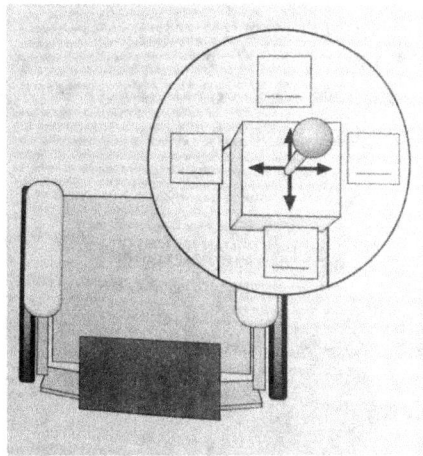

2. Here are electric range stove tops. Which burner is controlled by which knob? Why? (Affordances, Mappings, Cosmesis…)

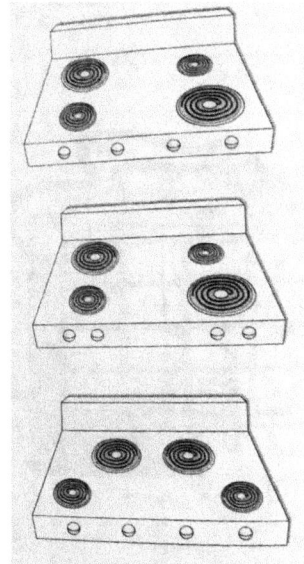

3. Here is an office prank on your keyboard. Your nearly finished report is due in minutes at the big quarterly meeting. What will you do? Why? (Knowledge in Head vs Knowledge in World, Ordered sets, Mapping, Feedback...)

4. Where on this child's older, durable 4 x 8, 32-item speech-output assistive device will you place these symbols or words: *under, on top of, down, up, over, left, right, around?* **What are your rationales for each placement?** (Mappings, Selectivity...)

5. This is the shower control in your hotel room. The bathroom is fogged up from the previous user, and you do not have your glasses on. Which way(s) will you move it for more water pressure...and for hot, cold, and off? Why? (Knowledge in World vs Knowledge in Head, Mapping, Selectivity, Transparency...)

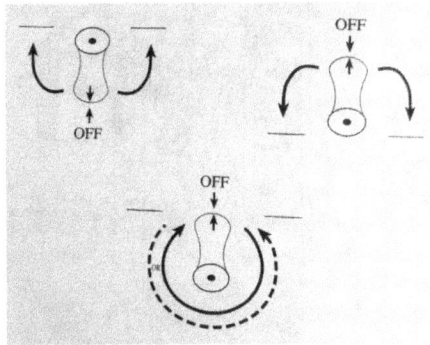

6. Your talented student is learning to play a musical keyboard device. You help her remember the keys by numbering each of her fingers. Which hand is *left* and which is *right*? What number will you assign to each finger. Why? (Mapping, Selectivity...)

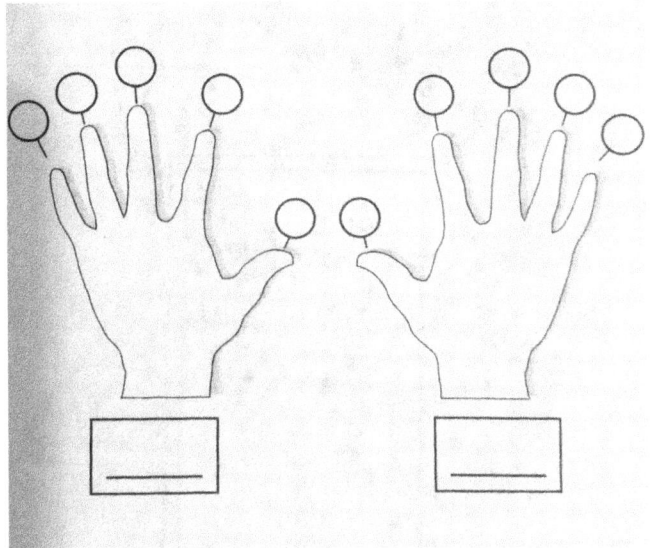

7. This is the door to your new, large storage closet. To open it, you will: Turn it up? Down? Pull it? Push it? Why? (Selectivity, Mapping, Affordances...)

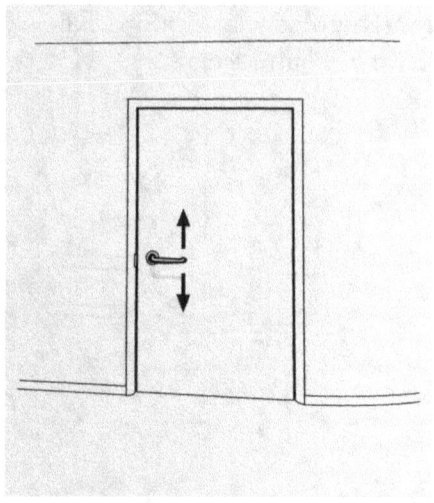

8. A local store displays bulk items this way. Costumers keep tossing in their candy wrappers and beverage cans. Why? (Affordances...)

9. Your volunteer group made nearly 200 pies of five different varieties for the community foodbank. Their pickup van is coming in an hour. How will you group the pies for efficient, accurate distribution? (Ordered vs Random sets, Chunking, Cosmesis, Mapping...)

10. This accessible washroom door at the café is advantageous for many users. Might some other users understand it differently? Why? How?
(Feedback, Mapping, Affordances, Sensitivity...)

REVIEW QUESTIONS

Beginning here, you will find a brief listing of questions and ideas to consider at the end of each chapter. Although not exhaustive, these questions can help you focus, integrate, apply, and remember information covered. Use them in any way you find helpful, and construct your own questions plus listings of topics and issues for further discussion and investigation as you read further.

1. Describe a device or tool totally new to you that was easy to use. Describe one that was difficult to figure out and to use the first few times. Explain why for each.
2. If you were to redesign the tools or devices above, what would you focus on so everyone could experience more success?
3. Do you or someone you know have a specific physical, sensory, or cognitive challenge that might affect how you/they might use those tools or devices above? Why? How? What characteristics of the items made users more/less successful? What could you do to redesign the items for greater success, safety, or satisfaction of use by everyone?
4. "People can't use stuff right because they're stupid, don't read the instructions, and don't pay attention." True? Not true? Sometimes? All the time? What do you think?
5. Wherever you are reading this right now, reach out and touch or look at an ingeniously designed tool, device, or technology--simple to complex--right in your midst: maybe a book, door handle, phone, faucet, light switch, other. How might the principles of design it exhibits apply elsewhere to other technologies.

2

YOU AND ME AND HUMAN FACTORS

Design is not just what it looks like. Design is how it works.
Steve Jobs

Several central human factors you encountered in the examples above deserve a closer look. Each of these concepts will be described and discussed more fully in other chapters. We will begin with these important areas, including some that you encountered in the preceding quiz. Several works of Anderson (2021), Helander (2006), Norman (1988 and 1993), Wells (2002) and others enabled this section.

If you have not taken the brief human factors quiz above, be sure to do so now to get you thinking. Some specific, essential human factors to consider in consumer devices and specialized assistive technologies include the following:

1. ***Transparency-Translucency-Opacity:*** overt, obvious visibility and user friendliness of the device to the user upon their first contact with it.
2. ***Ordered Sets Versus Random Sets:*** extent of groupings, organization, and predictability of linguistic and cognitive information needed to operate the device.

3. ***Cosmesis:*** look, appearance, and acceptability of a device regarding what other people will think of it.
4. ***Mappings:*** sequence and location of required movements and operations built into the device. How naturally or unnaturally does operation of the device match and respond to ways that we believe we should control or deploy it?
5. ***Affordances:*** what uses are suggested to us by the appearance, materials, and construction of devices with which we interact, plus how and what the design of the device or tool and the material(s) from which it is made communicate to us about how the device is to be used and by whom. Affordances can also convey impressions of viability, strength, durability, and inclusion for the device and user.
6. ***Learned Helplessness:*** potential users may decline their own attempts at acquiring independence, relying on family, care providers, or others, because it is easier and faster.
7. ***Feedback***: information provided by the devices used and especially by the switches and controls necessary to operate the items in question. Are there meaningful multi-sensory indicators that a user has successfully initiated or completed an operation (i.e., turned a device on or off, entered characters on screen, made a smooth turn in a wheelchair, saved a file, locked a car door...)
8. ***Knowledge needed to operate the device:*** based on what the user will likely already know, versus operational knowledge being based on something the user will have to determine or learn from the keypad, switches, controls, or screen of the device. "Knowledge in the Head" versus "Knowledge in the World" is discussed by King (1999), Norman (1988), Romich (1994), and others.
9. ***Constraints:*** the physical, semantic, cultural, logical, linguistic, or other cues are present with the device to indicate what you can and cannot do with it, and the way you are to do it (per Norman, 1988 and 1993).
10. ***Forcing Functions*** are built-in methods of backup and

functional safety measures assuring that mistakes or mishaps are prevented, or at least signaled minimized during device use (Interaction Design Association, 2021; Kantowitz & Sorkin, 1983; King, 1999; Norman, 1988 and 1993).

Inattention or, better yet, attention to these ten key human factors can have adverse or favorable impacts on success with tech. These parameters seem so obvious that we may forget how discouraging, even dangerous, ignorance of them can be. A few brief case examples of some essential human factors based on composites of actual consumers, clients and students will help illustrate. Names of all persons in this textbook have been changed to assure privacy.

---

100 words about *ORDERED SETS and CHUNKING*

Mnemonic techniques can aid our recall of location, movements, and information. Chunking refers to grouping similar things to make all in a set easier to remember. Ordered Sets are easier to recall and operationalize than Random Sets. The Memory Castle strategy of placing items to recall inside your imaginary castle halls, rooms or nooks enhances retention as you later stroll through them: an ancient way of retrieving ideas and facts. Your own cupboards may exhibit other applications. See https://www.toolshero.com/psychology/chunking www.animals.howstuffworks.com

---

CASE EXAMPLES

Transparency versus Opacity

Brent was seen at a local speech and hearing clinic for an assistive technology evaluation and appropriate intervention. A small, digitized speech output device was acquired for him, with the intent of sending it home with his parents each time to have them and him record some "news from my home" before each therapy session. Brent could then push the activation site on the device each time he came to therapy at the clinic to tell his clinicians his latest news from home— an excellent communication interaction idea, it seemed. His parents were taught how to erase and store messages on the device during a session and were given a copy of the quick-start instruction sheet on how to program the device in case they had trouble.

When they came with Brent for the next therapy session, they eagerly watched while the clinicians asked Brent if he had any news from home. Brent was excited to tell his news and pushed the appropriate activation site on the device. Nothing happened. The only sound from the device was background noise.

Brent was disappointed, as were his clinicians and parents, so another training time was held with the parents in the four-step process needed with this device to erase old messages stored at a given site and store new words spoken by his parents. The parents practiced in the session, took home the instruction sheet, and returned with Brent in a few days for his next session. It was "news from home" time, and Brent pushed the symbol site. Again, noise. Only noise, no speech.

Still another training session was held with the parents. Finally, at the next therapy session, Brent was able to report his home news. This technology application had validity and eventually worked, but the enabling communication tool that seemed so transparent in operation to clinicians (and to the parents when under the instructional supervision of the clinicians), proved to have too many steps and be too opaque for the parents to operate with their child when they were on their own at home. More parent training or a simpler device may have helped.

Mappings

Mary took a week off from work, and was staying at a quaint, beautiful bed-and-breakfast inn while on vacation. When she went to wash her hands for the first time at the sink in her room, she turned on the right-hand faucet, put her hands under the tap, and was nearly scalded badly.

She was used to turning on the standard right-hand side of the faucet handles for *cold* water (the usual mapping for these controls) and, without her glasses, could not see that the faucet handle markings were indeed reversed. HOT was on the right-hand side and COLD WAS on the left-hand side, an attempt to compensate for a plumber's error in hooking up the faucets when this older home was built.

Mary blamed herself for not noticing the reversal of faucet labels because of her eyes and wore her glasses during the rest of her stay. The inn's owner hired a plumber to reestablish the natural mappings for that faucet before others were hurt.

Affordances

Dean was 19 years old when he received significant brain trauma in an automobile accident. His resulting injuries left him with markedly dysarthric speech (slow and slurred) and with spastic quadriparesis of his limbs. For personal mobility he relied on an older nonmotorized wheelchair which required the assistance of someone to push. Because of his loss of motor and balance capabilities, he had to be strapped in tight. He accepted this reluctantly, saying he was depressed at the indignities of his restraint and the effort required by his attending helpers or family to secure him.

When Dean was about age thirty, he had an evaluation for adapted computer access. With his right thumb tipped downward, he laboriously typed "I am not a dog." He said he felt perpetually demeaned by the way people treated him with the wide leather straps and large, bare metal buckles used to secure him in his wheelchair. He stated these demeaned him, and that he felt treated like an animal in a harness with all the straps and buckles they put him in.

His consulting physical therapist and social worker explored getting newer, attractively colored nylon straps with color-matched clasps that looked more like standard seatbelts used in cars and planes. Even though they could not purchase a completely new wheelchair for him, they replaced Dean's rugged straps and bare buckles with softer, attractive, and more humane nylon straps with colored fasteners. Dean was pleased. While writing with his adapted computer, he typed that his revamped seating system made him feel less self-conscious and embarrassed, and made him feel "more like a human than an animal." The use of colored, softer nylon and attractive clasps afforded Dean a greater measure of human dignity.

Feedback from Switches

Roxy was 31 years old when she received her first tablet device. She had moderate-severe choreoathetosis, a type of cerebral palsy, and although she could speak, her speech was labored. With her newly acquired device, anything she could type on the keyboard, she could now speak through the device by using a text-to-speech app. When she was with others who had difficulty understanding her spoken words, she could clarify by typing in her message, then pressing the return key to speak it from the tablet. To aid her accuracy and speed of typing, she set the device to produce a classic typewriter key click each time she entered a character or touched the return/speak key. She and her listeners enjoyed the humorous old-time clicking from her high-tech tablet. Feedback from each keystroke, over type improved her typing speed and accuracy, especially when she needed to speak a synthesized message in fast-paced settings. A simple enhancement worked well for her.

As illustrated in the above examples, essential human factors utilized well or ignored either contributed to or detracted from these tech users' convenience, performance, inclusion, and safety across settings. Almost daily, we are remined by more publicized examples such as nuclear plant mishaps, space missions gone awry, plane crashes, train derailments, mistaken use of gun rather than taser, and numerous automobile accidents, that seemingly simple human factors and ergonomics issues play major roles in the safe effectiveness of technology as well as in its catastrophic failures. The examples of Brent, Mary, Dean and Roxy are different in scope and size than these larger disasters, but not different in type. The inability of technicians at Three Mile Island and Chernobyl to read warning indicators of overheating and then respond with correspondingly correct activation of the controls and switches to address the problem were significant human factors-related failures…

and true of the other, more personal yet potentially catastrophic examples cited above. The more smaller scale human factors-related experiences the individuals described above are no different or less important. They are humans attempting to interact successfully with machines and devices, and sometimes succeeding, sometimes failing. Their stories of device use, and ours and others', can be better understood through knowledge of and attention to practical, basic human factors ergonomics.

A USEFUL FRAMEWORK: BAKER'S BASIC ERGONOMIC EQUATION

An awareness of human factors can help product designers, educators, clinicians, and family members to understand what we need to know about humans and what to do with devices to make them more likely to interact successfully. Several major overarching factors combine to affect the likelihood of successful or unsuccessful human-machine interaction. These factors include motivation of the user; physical, cognitive, linguistic effort and load; and time spent in attempting to use a particular device or technology, as derived and adapted from Baker (1986), King (1999), and Westinghouse (circa 1880). A further detailed elaboration and interpretation of these factors per Baker's basic ergonomic equation (BBEE) will be found in Chapter 3. For now, sample overview narratives of these central factors in BBEE can help in understanding this complex, central relationship of motivation versus effort in success or failure with any tool, device or other technology.

$$\frac{\text{MOTIVATION TO USE AND SUCCEED WITH A TECHNOLOGY (M)}}{\text{PHYSICAL + COGNITIVE + SENSORY + LINGUISTIC + TIME LOADS (L)}} = \text{USER SUCCESS...or not (S)}$$

Motivation.

To illustrate further, first consider *motivation*. The user's desire to acquire, learn, and successfully employ a general or assistive technology appears as the primary factor in tech success. Spacey (2017) defines motivation as "the will to do things." In an ideal case, a potential user of devices grasps the need that they have and understands that others have been successful in using this technology for similar purposes. The consumer, student or client has observed others who are successfully using the technology in question so that they have models of what to emulate, and are willing to tolerate some frustration, failure, and fine-tuning of the system in pursuit of the desired system mastery (Anderson, 2021; Spacey, 2016 and 2017).

Cliché but true: nothing succeeds like success. A user who readily learns how to work the technology and can then become successful in its use in real-world increments, sets in motion a powerful, reinforcing snowball effect. The more you succeed at something, the more you want to do it. The more you do it, the better you get and the more success you have, and so on. Initial device transparency, the support of family and other partners in use of the technology in the real world, and carefully sequenced, age- and ability-level-appropriate teaching of the technology by clinicians, educators, and other interaction partners can all combine to set in motion this powerful cyclone of success that intensifies with each use-success-use-success revolution.

Perhaps the culminating portion of motivation can be best summarized as normalized use of a device or tool in the real world. For example, when your adapted mobility system gets you to and from work, or when your communication enhancements truly allow you to converse and to interact with kids on the playground in ways that are more like what other kids do, then that is the epitome of self-perpetuating successful use of tech—and the primary catalyst of durable motivation.

Other human factors that can have a dramatic impact on successful use might be grouped together as *load* or effort factors. Load means how much nuisance, encumbrance, weight, burden, or effort demands a device or method puts on the user during implementation. *Exertion* is another way of expressing the concept of load, work, or effort expended by a human in accomplishing a task with a device. It is a variable that we wish to reduce as much as possible in our tech interventions. For our purposes here, *load* and *effort* will be considered operationally synonymous. Component areas of load or effort consist of these five major divisions: *physical, cognitive, sensory,* and *linguistic* effort, plus *time* to complete a task. An applied overview of each component helps understanding the combined load factors in tech.

Physical Effort

Physical effort refers to how much muscle movement and exertion are required to initiate, pursue, and complete a task. For example, "muscling" a wheelchair up a steep slope to get to the front doorway of the building in which she has a job interview may simply be too physically demanding for a given user. She tries but cannot generate enough force to move up the slope, even though she is highly motivated to do so. In another example, a swimmer tries repeatedly to climb a boarding ladder back on to a boat after diving into the water and cannot hang on in the rocking waves. He gives up climbing because it is too much physical work, and yells for help. We have all experienced physical load factors in the common tools and devices we use every day, from ladders to jar lids to bicycles pedaling up a hill. Some tools and devices require more effort to use. Given a choice, we tend to use, reuse, and develop skill with things over time that will require less effort, less physical exertion. Simply put, we all tend to use devices that are easy and avoid technologies that are not.

Cognitive Effort

In addition to tools and tech being physically challenging, sometimes the *cognitive effort* (thinking) involved in their use constitutes a heavy load. Cognitive effort factors include sensing, remembering, discriminating, analyzing, problem solving, and sequencing of the actions and responses needed to operate tools and devices. For example, one of the reasons that many people do not program the clocks and timers on their home appliances or entertainment tech is that it requires more cognitive effort to decipher the instructions than they are willing to expend. In another example, our pocket smart phones offer an array of features that often go untapped. It takes cognitive effort to get the most out of them unless the device automatically sets up the features and makes them easy to access. As with other familiar devices, if cognitive demands are too pressing, especially over repeated tries, we tend to give up our attempts to use the device in more elaborate ways and decide to just stick with the functional basics, not the frills or add-on features.

Sensory Effort

Visual, auditory, tactile, olfactory, haptic, stereognosis, balance and other senses may be required for successful use of devices. Perception, acuity, discrimination, and judgments across senses add load factors we often forget. Consider these technologies which require sensory effort to use well: cooking stoves, entertainment equipment, fireplaces and wood stoves, refrigeration, home or business natural gas leaks, water and sewer hookups, home and car heating and cooling, as well as enabling devices across senses that may mitigate usage.

Linguistic Effort

A close relative of cognitive effort, *linguistic effort* refers to the amount of reading, perhaps in other languages, symbol interpretation and icon processing that the user must invest to operate a device. For example, vending machines for soft drinks generally have a full color logo of the type of beverage placed prominently on the selection buttons. Users merely match the symbol, logo, or color associated with the type of beverage they want, rather than reading the actual names of each on the selection switches. The typed or printed names and descriptions would work for many users as a way of designating the individual items to be selected. But use of the logo or symbol plus related colors reduces linguistic load on a wider variety of users (who may not be able to or will not read), allowing for accurate selections of the beverages they wish to purchase. Linguistic load is also affected by the sequencing of symbols that must take place to operate some devices. For example, doing a "control-alt-delete" or other split-finger function sequence on a standard PC keyboard requires more recognition of symbols in sequence than does accomplishing similar screen commands by using the mouse, trackpad, or other pointing device, wherein the user simply moves the pointer to the desired selection buttons or icons one at a time and then activates the appropriate side of the mouse buttons. Linguistic load becomes greater as more reading, listening, interpretation, and sequencing of iconic visual symbols, words, sound signals, and other encoded forms of information are used in accessing, operating, and controlling a device.

Time Load

The culminating effort factor, which expresses the sum of all the other load factors, is *time*. The amount of time needed to accomplish a task using an assistive or consumer technology is a function of physical, cognitive, sensory, and linguistic, effort factors combined, or it may reflect the real or perceived complexity of the task required. For example, one of the reasons that most people probably do not write postal letters as much now as in earlier years is not that it is too physically, cognitively, sensorily or linguistically exerting, but that the sum of those takes more time than they wish to spend on the task. If an activity proves to require more time than we wish to spend on it, we often abandon it in mid-task and probably will not resume it unless our motivation to do so drastically increases. The time to pursue and complete a task with a device also can be reflected in the type of special need, limitation or disability presented by the user. Persons with severe spasticity or athetosis, vision impairments, or limb differences may find that some important tasks, such as typing on a standard keyboard, operating a phone, tying shoes, fastening clothing, or preparing meals take too much time to accomplish in standard ways. When these tech users investigate assistive strategies and technologies to help them in these tasks, they primarily want devices and methods that will make their work in these areas faster. They may well be able to accomplish these tasks without AT, so the physical, cognitive, sensory, or linguistic loads may not be prohibitively high, but the main barrier is time on task. If we can do a task quickly, we will likely choose that activity over something else that may not be harder, but which takes more time. Hence, instant oatmeal, microwave dinners, and other time savers are abundant, and are often used in the general consumer marketplace, as well as the many hardware and software innovations across all component areas that make using tech faster...thus easier. To enhance understanding, here are some brief case examples composited from real students and clients to again highlight each of the factors described previously.

CASE EXAMPLES

Motivation to Pursue and Complete a Task

Chad was 11 years old when we knew him. He had cerebral palsy, manifested in a severe right diplegia that markedly limited his use of his right hand, arm, and leg. Chad was the first born of four brothers, all about one or two years apart in age. He had never shown any interest in sports or physical activity until two of his younger brothers started playing soccer with a new local youth team. Chad soon became motivated to be involved with his brothers; he wanted to play with them and to be like them. He began playing "sit-down goalie," as the boys called him, with his brothers during backyard games. Chad found that he was able to move better and in different ways when he was relaxed and having fun than he and his family had ever thought possible. Although he could not stand or walk without support, he was able to play goalie with his brothers at a level that they considered satisfactory for fun. He was motivated to be included as "one of the guys," and he achieved movement capabilities and derived enjoyment from movement that his previous years of physical therapy and adapted physical education had not yet succeeded in bringing to him. He was highly motivated to be involved and was willing to tolerate and even overcome some movement limitations, awkwardness, and physical pain to play with his brothers.

Physical Effort

At age 78, Delores had severe arthritis and lived alone in a senior apartment building in her city. She found that her mobility around her apartment, her neighborhood, and her city were limited because of her physical condition, so she relied on her rotary dial phone to order groceries and prescription medications, and to keep in touch with her daughter, her friends in this town, and her sister in another city. She made frequent calls to them every day. Eventually, the physical effort and pain involved in dialing her old rotary phone became too much for her, so she stopped calling years ago. Her daughter and sister became concerned at her isolation. After discussing the matter with Delores, tried to find a home phone that was easier to use. Delores's daughter went to a local discount store and purchased an off-the-shelf phone with large activation sites (buttons) and speed-dial features. Delores could now still dial her most frequently called friends and relatives from home, as well as emergency numbers, with just one push of a big button on her phone. Her phone use eventually returned to and exceeded previous levels because she did not have to invest as much physical effort (and pain) in the repetitive task of dialing; it was simply now much easier for her to do. Her family also set her up a mobile smart phone that had large activation sites on the screen and large, clear letters and numbers she could read in all types of lighting, indoors or out.

Cognitive Load

Billy, a boy of 7, was functioning at a much younger level. He was presented with a rocker switch to control two switch-operated toys. Separating the functions of the left and right sides of the rocker switch and the toys on the respective sides of the table that they controlled proved too difficult for Billy. To lessen his cognitive load and make the situation easier to understand, each toy was connected to a separate switch, with each switch having a colored marker as well as an individually distinct tactile landmark. Billy was more readily able to differentiate and activate the toys individually and at will when the activation sites were separated and made sensorily more distinct from each other. This configuration of switches put less cognitive load on Billy improved his motivation, performance, and enjoyment of play.

Linguistic Load

Kendal, age 23, was in her second year of university work, majoring in journalism and communication. She was successfully living with mild athetoid cerebral palsy and learning disabilities, and though not an outstanding student, she coped with the demands of school by working hard and putting in long, laborious hours in her studies. She was doing well, all things considered. But now, as a university junior, she found that the demands on her to write many papers and articles, all with near-perfect grammar and spelling, were increasing. These demands for more writing were taxing her abilities, stamina, and motivation to finish her degree. In consultation with a local tech clinic, she learned to use the built-in sticky-keys and other adapted access features on her PC, phone and tablet. These aids allowed her to do multi-finger functions in sequence rather that all at once, as she became skilled at relying on word prediction and copy-cut-and-paste features to help her reduce her numbers of keystrokes. Kendal's student internship at a newspaper office involved covering the police and court beats. These topics, with their special vocabulary and often repetitive use of terms, lent themselves well to the use of word prediction in her writing. Kendal used built-in word processing software that included a thorough, large spell-checking program as well as grammar-checking functions. These highlighted her possible mistakes as she wrote on screen and allowed her to fix them before her copy went out for editing. Kendal's productivity and confidence increased as a student and journalist.

Time

Andrea, a literate woman of 38 with mild developmental physical and cognitive disabilities, also had a closed-head injury in a car accident at age 23. She recently gained employment at the county courthouse, helping to do general cleaning, collection of recyclable materials, and emptying of trash containers for several floors daily. Her neatness, accuracy, dedication, and overall attention to detail and thoroughness in her work were remarkably good, even though her pace of working was slow. Her other attributes—punctuality, dependability, and willingness to learn—made her a valued, trusted employee. Her main concern about her job was in having to use the elevator independently to get to and from her work on the higher floors of the office building. The elevator doors opened and closed faster than she could maneuver herself and her walker in and out. She feared the doors closing on her, as well as the embarrassment of getting hit by them or needing to call for help with the elevator. She expressed these concerns to her supervisor after a few months, and the elevator company was contacted. They adjusted and slowed the rate of door closing to match Andrea's abilities more closely, and she has used the elevator on her own since, with more confidence and less interfering anxiety.

---

100 words about *WHO ARE HUMANS?*

We current humans are *Homo Sapiens*, "wise man." Extinct lines of our global human family include *Neanderthalsis* (Neanderthals), *Denisovensis* (Denisovens), *Floresiensis* (Floresiens), and *Luzonensis* (Luzoniens), plus *Homo habilis*, *Homo rudolfensis*, *Homo erectus*, *Homo antecessor*, *Homo heidelbergensis*, and *Homo naledi*. Others? Human factors and ergonomics considerations not unlike many of ours today were likely encountered by our ancestors. Much of what we know now about past humans derives from their tools, art and tech. See https://www.nhm.ac.uk/discover/the-origin-of-our-species.html **and** https://www.livescience.com/how-many-human-species.html

---

SUMMARY

Attention to essential human factors helped foster success in all these cases. The technologies themselves are readily available and modifiable. What are more often not considered are the human factors and related barriers that must be identified and dealt with to make the technologies and their users truly unified, viable, successful systems, working together. As we address the goals of practice, we will continue to expand our view of what human factors are and how they may be included in our efforts to develop successful accommodations and interventions. Across component areas and variations of human abilities and challenges, human factors considerations contribute to effective, successful, and safe use of tools, devices, and other technologies.

REVIEW QUESTIONS

1. Name four common tools, devices, or technologies which you have found to be easy or pleasant to use, and four items you have found hard or unpleasant. Why?
2. Which items of the brief human factors quiz early in this chapter seemed most obvious to you? Which seemed puzzling? Why? What are your experiences?
3. Describe a consumer, student, or client you have known who has succeeded or failed in using tech of any type. How about you? Was success or failure related to one or more human factors areas we have mentioned so far? Which? Why?
4. What are the five areas mentioned and depicted in Baker's equation? Name and define each of them as you diagram them.
5. Describe a situation in which you or one of your customers, students or clients found that motivation to use a device was overwhelmed by failure with one or more of the effort factors we discussed. Did you or they try to use the device or do the task that way ever again? Why or why not?
6. Describe three areas of human factors that are addressed well in general consumer technologies. What about the same for AT? Describe three examples of human factors violations you have observed in products. Why in each case?
7. Apply Baker's Basic Ergonomic Equation (BBEE) to a technology from each of several areas in your own life (communication, computer use, mobility, environmental control, etc.) Will considerations expressed in BBEE affect how much, how often, or how well you use those technologies? Why and how?
8. Which of the human factors concepts and ideas described above would you disagree with? Which may have changed or stayed the same for people over time? Why?

3
DESIGNING FOR PEOPLE

You cannot understand good design if you do not understand people. Design is made for people.
Dieter Rams

Any sufficiently advanced technology is indistinguishable from magic.
Isaac Azimov

It is our nature as humans to use tools. Although use of language and complex communication by human beings are often cited as primary variables separating *Homo sapiens* from other intelligent species, it may be just as true that we are the only users, inventors, and refiners of complex tools for a multitude of uses in the many environments we inhabit.

Crows, beavers, chimpanzees, and other animals are also users of natural tools for immediate tasks. To date, however, none have been observed fashioning tools from refined ores, or revising and adding mechanical, hydraulic, or electronic enhancements, then making and distributing devices on a large scale throughout a given animal society. Could that happen? I hope we observe it. There is a lot we have not yet noticed about other tool-user species.

Although we hear of smart birds who can unlock their cages, clever dogs and cats who can manipulate their entry doors or food dispensers, and even cunning velociraptors (just in the movies) who can thwart most human-designed barriers, it is humans who have used, perfected, and passed along complex, refined tools, and tool-related processes to assist in our lives and work. Humans have done that for thousands of years. We are good at it.

Humans are not the only creatures aware that technologies help. Tool use is reported among animals, birds, and insects.

Even people in what some may still consider primitive or aboriginal societies, both past and present, used and are using still impressive arrays of sometimes simple, sometimes deceptively complex technologies. Their tools and devices assist them in accomplishing the tasks important in their daily lives per Stone (2021) and Washburn (1960) among others. Food gathering is surely advanced by the design skill and mechanical technologies employed in fashioning a reed basket, spear, or bow. Canoes, paddles and harvesting sticks have been refined in North American gathering of wild rice, the food that grows on water. Worldwide, other effective indigenous technologies for surviving and thriving have developed with stellar ingenuity over the ages of diverse humans.

Additionally, hunting, and one's likelihood of feeding a family or community regularly, can be enhanced in some areas of the world by use of chemical technologies, poisons, derived from natural sources to increase the effectiveness of an arrow or spear. Chemical technologies also have been used for thousands of years in the cloth dyes and pharmaceuticals developed by purportedly primitive societies worldwide. Sewing needles made from bone; thread made from animal sinew or plant fiber; portable, secure housing made from saplings and hides; watercraft made of materials as diverse as walrus skins, birch bark, or modified tree trunks ... all these technological innovations accomplished in the natural, "primitive" world indicate how resourceful we humans truly are in learning to adapt to, move about in, and engineer our diverse global environments.

This resourcefulness is one of the main reasons that we as a species have proliferated so on this planet. Survival may indeed be for the fittest, particularly when "fittest" is defined as those who can fit into challenges of their environment by adapting to and gaining from it by use of tools and technological processes.

It can also be argued that one of the main characteristics setting humans apart from other intelligent creatures is our capacity across cultures for providing long-term nurturing and support for our growing children over extended time. And perhaps most notably for our purposes here is our singular concern and care shown for the disabled members of our human family. At least in many past and present human societies, and most observably in current Western culture, a shining characteristic of civilization, whatever our other human faults, is our awakening commitment to care for and enhance the lives of people with special needs and challenges. Moral, ethical, and legal sanctions in many countries attempt to promote the humane inclusion of disabled persons in growing aspects of daily living, education, recreation, and employment. A worldwide enlightenment regarding disability is occurring, but there is still much room for progress.

Specialized accommodations for inclusion of persons of all ages with numerous special needs in the daily life of a society are bringing out the best of human creativity and innovation. Equipment, devices, and other tools to help meet the specialized needs of our very young, disabled, or very old societal members have become increasingly recognized, accepted, and important. These innovations can enhance life in all its aspects for someone with special needs by conferring a measure of independence, autonomy, and productivity that would not otherwise be possible. Imagine teaching children who have moderate-severe hearing loss without use of any sound-amplification devices. It can be done, but such practice would be a substantial, detrimental departure from methods that have become essentially standard in Western society over the past hundred years. Imagine always having to carry or bodily support an adult who has difficulty walking, but who could do so alone if a cane or crutches were available. Such seemingly simple uses of tools and technological processes are, unfortunately, still not globally available to all who need them, even though they are unremarkably common in industrialized societies. However, for those to whom these basic assistive devices are available, the world is changed. The simplest, sometimes most commonplace applications of technology in the lives of persons with special needs can make dramatic, irreplaceable differences in the quality of their lives—as well as in the lives of their families and others who care for, educate, and work with them.

Human creativity fortunately continues to rise to the challenges posed by special human needs. The development and use of assistive, helping devices and equipment for people who have special needs or disabilities is the field of *assistive technology* (AT), also known as *enabling technology*. Both terms convey the concepts of tool use and tool-related processes applied specifically to allow people with special needs, disabilities, and challenges to participate in as much of life as possible along with others in their family, community, school, and worksite.

Lloyd, Fuller, and Arvidson (1997), Church and Glennen (1992), and others have described component areas of AT that include: augmentative and alternative communication (AAC), adapted computer access, assistive listening and seeing devices, environmental control (EC), adaptive play, positioning/seating, powered mobility, and integration of technology into the home, school, and community for worksite access and modification, among others. Assistive or enabling technologies are often thought of as being overly complex, microprocessor-based powered wheelchairs, or elaborate speech-output computer systems, and they can indeed include these things. But more often, assistive technology may mean a slightly bent spoon so someone with a wrist flexion limitation can eat independently, or maybe a lowered drinking fountain so that it can be used by a person in a wheelchair. Useful enabling technologies do not necessarily have to be electronic, expensive, or complex.

More formally, assistive technology was defined in U.S. federal statutes. AT is described by the Tech Act of 1988 (Public Law 100-407, the Technology-Related Assistance Act) as including "any item, piece of equipment, or product system, whether acquired commercially off the shelf, modified, or customized, that is used to increase, maintain, or improve functional capabilities of individuals with disabilities" (Section 3.1). Public Law 101-476, the Individuals with Disabilities Education Act of 1990 (IDEA), relies on this same definition but uses the term *children* in places where the Tech Act of 1988 uses the word *individuals*. Federal law and guidelines in the United States, as well as in individual states, have been updated, and are in continual revision and legal interpretation since. Society has kept evolving toward greater inclusion, but that is not a guarantee. Constant vigilance of inclusion of all people, and acceptance of differences is always needed.

Other nations or cultures may not rely on these U.S. definitions, but these are a basis from which to better understand the nature of AT. Overall, academic study in and clinical practice of AT are complex, eclectic fields. *The range of what can be assistive technologies varies from complex and highly specialized to seemingly simple and mundane.* We must remain alert and accepting of all.

Indeed, the most common, formally prescribed assistive technology relied on by well over half of the population of North America is optical: glasses and contact lenses. Assistive, enabling optical technologies have become so commonplace in the United States, Canada, and most of the Western world over the past two centuries that we take this effective, inexpensive, life-enhancing technology mostly for granted. It is a tribute to the effectiveness of these optical technologies, their inventors, designers, and clinicians, that glasses and contact lenses have become almost literally an invisible assistive technology in our society.

We seldom think about whether someone may be using AT for his or her eyes. In many cases we cannot even tell at all that the person is using lenses. The acceptance of glasses and contact lenses seems to be virtually universal in our culture; there appears to be little or no discernible social penalty for wearing lenses or using glasses, regardless of age group. We might say appropriately, with only moderate pun intended, that the use of assistive optical technologies has become transparent.

BENNY: An instructive case example.

Benny, age five, had come to a regional rehabilitation clinic for several evaluation visits. Benny had significant difficulties with speaking due to the severe motoric limitations imposed by cerebral palsy, but he also seemed to be a bright boy who understood well, according to his parents and teachers. He had much to say and the intent to say it, if only he could talk.

After several hours of careful assessment and evaluation with Benny over a two-week period in late July, his teachers, therapists, and assistive technology specialists met with his parents and recommended a type of specialized voice output communication aid (VOCA) for him. Benny had used this type of device with encouraging results in trial runs at the clinic, and it appeared that this device would finally give him a voice. It appeared that Benny soon would be able to "talk" on his own by learning to use his VOCA in support of his own oral speech, which was still developing.

His parents were cautiously pleased. Together with Benny's teachers, therapists, and other support personnel, they had worked as a team to make careful, well-reasoned decisions about the type of assistive communication technology to acquire for Benny. They initially used a rental device, and in several weeks their application for funding and purchase of the device recommended for Benny was approved by his family's insurance provider. The device arrived in early October, and instruction and practice with it began on a regular and intensive basis for Benny. His parents were excited to use it with him.

About three months later, after the Christmas break, Benny's teachers and therapists at his school noticed that Benny was not bringing the VOCA to school anymore. They thought he and his parents might simply have forgotten to bring it, so they phoned Benny's parents to remind them to be certain to send the device with him each day on the bus now that school had started again.

The parents responded that over the course of the holiday break, they and Benny had done much thinking and had decided that the speaking device was too hard to use. They were tired of programming it, they said, and besides, Benny was tired of lugging it around. They also said that its voice sounded funny to all their relatives, who ignored voices that came from the VOCA because it wasn't Benny's real speech, and they did not need the device for Benny after all.

Benny's parents said that they had realized over the holiday that he communicated fine (in their opinion, *better*) with his family and relatives without the VOCA. They did not see any reason for Benny to keep dragging it around with him, for at his age he might even break it, and it had cost so much that they did not want that to happen. His parents concluded that, overall, the speech-output device had been fun for Benny for a while but had now become a nuisance. They reported that it was safe and stored under Benny's bed for when he got older.

ENTER HUMAN FACTORS

An unusual story? Unfortunately not. A growing body of shared clinical experience and documentation among developers, manufacturers, teachers, therapists, clinicians, and others who work in the field of assistive technology suggests that this is a common scenario. In my own experience as a speech-language pathologist and special educator for four decades, I have witnessed it many times. Sometimes, despite our best, most learned clinical efforts to select, prescribe, and acquire assistive devices and technologies for our students, clients, and patients, there seems to be some additional, critically important but missing elements for which we must account. Elements that have a significant impact on the likelihood of success or failure of our AT intervention efforts with persons of all ages and disability types. These important variables are often ignored or not even considered in our AT practice. They are known as human factors.

So what is new about human factors? Isn't that all obvious? As alert consumers, teachers, clinicians, and therapists, we know a lot about humans. We are trained in the diverse, complex clinical aspects of assessment evaluation, diag9nosis, and intervention with our clients. We in these helping professions are also receiving better training now than ever before in the entire assistive technology area for professionals at the preservice (university) and the in-service (continuing professional education) levels. But even amid our rigorous professional training across many related fields, and amid our own consumer and clinical experiences, we continue to encounter these other vexing human factors: variables often left unmentioned and unaccounted for that define the use of tech by us, our customers, students, clients, and patients.

Despite our best efforts to get people the assistive technology they need, and to help them learn to use it, the AT user and his or her family may give up or refuse to use it. We tend to regard this as a clinical failure of theirs or ours, often for reasons that we may not understand. But this rejection of tools, devices, and technologies may be more related to inattention to essential human factors in AT device design, selection, and use. Because of the rising costs of health care, education, and specialized services and devices of all types, acceptance and continued successful device interaction are becoming increasingly major considerations in our work with persons for whom we recommend assistive technology. Rejection and non-use of costly AT devices waste critical resources and time. Device rejection can occur despite our best clinical efforts, and it may occur because we overlook the seemingly obvious essential human factors involved. I have seen this happen often. You may have experienced it also. Our ignorance of, or inattention to some basic, essential human factors (that may seem on the surface to be just aspects of common sense) can negatively affect our clinical practice in AT. Also, overt, concerted attention to essential human factors may mean the difference between success and failure over time for many of our potential AT users. An injection of practical clinical wisdom into a growing emphasis on clinical and technical skills is necessary. That is the reason for this book of basic, useful concepts in human factors.

Practical applications of human factors in assistive technology, or any other area of technology, are content areas in which we, as teachers, therapists, and clinicians, are typically not educated and trained.

Most often we have had to learn about their existence and importance through clinical experience, some of which has been bitter, and much of which has probably included too many failures of our intervention attempts that were otherwise well devised and intended. To confound the problem, some professionals or others still may be unaware of the existence or role of human factors. They may even consider human factors in consumer and assistive technologies to be trivial or merely in the realm of common sense, or to be such basic, intuitively apparent considerations that these are obvious to any educated professional, and not worthy of our professional consideration.

And that is *not* accurate. The complex field of human factors and ergonomics in engineering and design for general consumer technologies and products is far more than just an iteration of the obvious. So too are human factors in assistive technology. As an effort to introduce and discuss some essential human factors in assistive technology, this book will help shed light on this important topic.

Human factors in design and engineering of general consumer devices and more specialized assistive technologies is a complex field, and this book is not an engineering text written by a human factors engineer. It is a clinically oriented, practical, and readable effort by a speech-language pathologist and special educator who has worked with assistive technology for many years; a clinician and family member who has encountered many human factors-related successes and failures with numerous people living with special needs. My practical interest and involvement in human factors and ergonomics in applied tech began decades ago, but were never addressed directly by any text yet in my clinical or assistive technology areas. My learning therefore has had to come from a variety of sources, colleagues, clients, and life experiences with family. I recognize and thank the many clients, students, parents, willing families, and professionals who have taught me so much that is not yet contained in journals or books over these past years of clinical practice.

More recently, the dwellers of the human factors and assistive technology on-line sources have provided insights, challenges, discussion, and wisdom from diverse tech user's points of view. The chats and arguments we have online or in person, and the stories and experiences that many AT and other tech users and families share are consistently valuable in the development of this book.

In addition, much credit for my and others' awakening and learning about human factors were due to these main sources. Over the years, their central ideas, concepts, and information applied and interpreted have helped me think about, expand, and utilize apply their human factors wisdom in my own life and clinical practice and :

A Guide to Human Factors and Ergonomics by Helander (2006)
Human Factors in Design and Engineering by Sanders and McCormick (1993)
The Psychology of Everyday Things by Donald A. Norman (1988 and 1993)
Human Factors: Understanding People-System Relationships by Barry H. Kantowitz and Robert D. Sorkin (1983)

These and other essential books, online sources and many related experiences, client contacts, conferences, and family needs influenced my thinking and clinical practice. I have extrapolated and applied ideas contained in them to my work and life in a variety of ways over the years, and enthusiastically encourage you to explore and consult these authoritative sources for further depth and detail beyond the scope of this more clinical, less engineering-oriented book. You are also encouraged to consult the numerous other excellent references on which this book is based. There are many, and this text would not exist without them. Full bibliographical citations of all are made in the reference section.

Back to this current book. As extended and synthesized from these sources, as well as from many others, **ADD HUMAN** has several enhanced purposes. It is intended first to underscore and highlight what you and I may already know about human factors in across different technologies, helping us recognize and apply it in our lives, work, and clinical practices. Another purpose of this book is to alert us what we still need to know to enhance our work. Many missing pieces remain.

In addition to these purposes, and perhaps even more important, this book is my effort, as a family member of persons with special needs, and as a working clinician and educator, now retired, to stimulate you, me, and our colleagues in a variety of fields to assess continually what we *know* and what we *notice* about the world of tools, devices, technology, and about how persons with special needs and limitations interact with them. This book is also a formal effort to cause us all to continue to recognize what else we need to learn about tools, technologies, and the design, selection, and use of them in the real world as rapid tech developments continue, and from the perspective of those who have special needs and limitations. Overall, this book is intended to help us better understand the perspective and needs of those who may approach and interact with devices from a completely different point of view and experience repertoire than yours and mine. Most persons with normal-range abilities to interact with a variety of devices, tools, and technologies do this so often, so routinely, and so well that we may never consider that someone else from a different experience base, or none, may view similar interactions quite differently. What is obvious and intuitive for many of us may not be so for others.

Some essential human factors in assistive technology may be viewed as obvious and common sense, but we disagree. We will list and describe some central human factors in assistive technology in ways that readers from a variety of levels of preparation and experience, can understand and use. I hope you will find helpful applications in your work with your students, clients, and patients as they learn to use the tools and technologies you introduce into their lives. Helping you to help them through your increased awareness, through your application of existent knowledge, and through serving as a catalyst for the stimulation of further discussion, research, and refinement of our understanding of essential human factors in assistive technology is the purpose of this book.

TECH TRANSPARENCY...AND MORE

In general, people prefer to use things that are easy and nonthreatening to operate. Therefore, the best assistive technologies, we would argue, are in fact *transparent*. *Transparency* is a term borrowed from the field of augmentative and alternative communication (AAC), a component area of assistive technology that we will address later in this book. In its strict definition, transparency in the realm of AAC refers to how evident the meaning of an icon or sign is in a nonspeech symbol set or system as described by Fuller and Lloyd (1992), Silverman (1995), Von Tetzchner and Martinsen (1992), and others. Signs or symbols that are readily guessable, that is conveying meaning in an open way that virtually any viewer can understand, are said to be transparent.

This same concept might also be generalized to assistive technology itself. The more obvious the method of use or operation of a given device or system, the more transparent the technology is to the user. Norman (1988 and 1993), in his illuminating texts on common technologies and tools refers to this concept as visibility. Consider common everyday items such as door handles, household light switches, or faucets. Most are transparent, visible tools and devices. Their operation is largely self-evident, allowing users to accomplish a variety of important tasks without special training, and many of these items are usable by persons of varying abilities. The function of each is generally so transparent, obvious, and transferable to other devices of the same type, that if you know how to work one light switch or door handle in a culture or place, you can probably intuit operation of most other light switches, door handles and so forth.

Transparency appears desirable across most consumer products and devices. Overall, the best consumer technologies are designed and tested to be as transparent, as visible as possible for the general user. It could be similarly argued that the best assistive and enabling products, tools, or technologies intended for use by persons with special needs should likewise be as transparent as possible for people with the widest possible range of abilities, including the technology user plus his or her family, care providers, teachers, and clinicians. Overall, it seems that the more evident and easier a tool is to use, the more use it will get by all.

Translucency is another term used by Fuller and Lloyd (1992), Silverman (1995), Von Tetzchner and Martinsen (1992), and many other authors in the AAC field. It refers to symbol vocabulary items that are more guessable when one understands some of the background or associative meaning connected with the sign or symbol. The icon may not be transparent enough to be readily comprehended on its own merit, but when the design rationale of the sign or symbol system is known, then meaning can be more readily discerned.

Similarly, it could be said that certain technologies such as many consumer electronics, smart phones, tablets, wrist monitors and watches, computers, digital alarm clocks, ovens, and others tend to be translucent in operation. To many users it may not be readily evident just how to operate the device unless you have been shown how, or unless you have taken the time to read or receive special instruction in operation specifics. Many if not most of the more complex assistive technologies with which I have had clinical experience are best categorized as *translucent*. They are like other technologies in our experience repertoire, but they require some special individualized instruction, configuration, or perhaps detailed preparation to use properly. Thus they are translucent rather than visible or transparent. Although we can discern their proper use eventually, their use is not readily apparent upon our first, uninitiated contact with the device.

Another relevant categorical concept used in AAC, as described by the authors and others is *opaqueness* or *opacity*. In symbol or sign sets and systems, an opaque item is one whose meaning is impossible to know unless you are specifically taught it. A set of defined, specialized referents or instructions must be provided to discern how to use and interpret the sign or symbol. Assistive technologies, devices, and systems of devices can also, we would argue, be *opaque* to users, families, and professionals. For example, even a common transition from one basic computer operating system to another, such as a change from Macintosh or Windows platforms (or vice versa), can interfere with our productivity. This type of transition, which necessitates new learning of an initially opaque system, can be daunting to some technology users and can cause them to reduce or postpone use of the computer (Stoll, 1995; Rathbone, 1995). In my experience, some of the more complex, microprocessor-based adapted computer access devices for adapted communication or environmental control can be difficult to configure and operate and may be frustratingly opaque even though they are intended to accomplish such basic common tasks as word processing, calculation, or phone calls. Some assistive, enabling technologies that appear simple and seem to have promise in the catalog description are often so opaque to the less trained user, teacher, or clinician that they are rendered effectively useless in the applied environment of home, school, or clinic. What is transparent to one user may be translucent or opaque to another with differing experiences, technology, training, or interests.

In the real world of clinical practice and daily life of AT use, the more opaque a system is, the less likely it will be used to the extent or duration that its potential warrants. In short, we tend not to use a device well or for long if the device is too difficult to figure out or to set up each time we want to use it.

For example, Ross Thomson (1988) and Angelo, Jones, and Kokoska (1995; Angelo, Kokoska, & Jones, 1996), among others, have alluded to or reported that after a few months of initial enthusiasm, some parents of children living with severe disabilities may store away and not use a complex AT device or system that is found to be too complex or difficult to use in the home, even when the initial acquisition costs, time, and effort were high. Parents in these and other studies have indicated that the use of some AT devices was just too much trouble for the home environment. The device may have taken too much time and effort to program and reprogram as their child tried to use and develop skill with it. Thinks of the Benny story earlier. Overall, some devices seemed too intimidating technologically; that is, access to them was too opaque to encourage parent comfort and continued use.

How many AT devices and systems of all types have ended up in a storeroom because they were too opaque for their users? How many AT items are now under the bed or in the storeroom because they intimidated the user and his or her family? On the other hand, some specialized technologies must be complex to accomplish the unique purposes for which they are intended. Perhaps the true challenge lies in making the AT device or system capable of accomplishing its intended work, but keeping it user friendly enough to be at least translucent, not opaque.

It appears that for all devices and systems across all component areas of AT, a primary, essential human factors reality may be the relative transparency-translucency-opacity of a device for a given user. Even when considerable initial enthusiasm surrounds the acquisition of AT, if the user, family, and other care providers are ultimately intimidated by its utilization and configuration, the AT system will not be used to its full potential. Overall, ease and visibility of operation for assistive technologies are fundamental, basic human factors considerations in successful AT intervention. Full AT transparency is most desirable. AT translucency is acceptable and most common. Opacity portends user failure.

---

100 words about perils of *TRANSPARENCY AND VISIBILITY*

Our neighbor Vince, mid-seventies and an experienced outdoorsman, heated his home with wood. Working in his large backyard, he cut up logs and split wood into stove-sized pieces, taking a large pot of hot coffee with him most mornings. Vince lamented that his coffee got cold and sometimes froze before he could drink it on frigid days. We surprised him with an insulated dispensing jug featuring a simple lever pump. Weeks later, I stopped by to chat, noticing he unclamped and removed the entire jug top assembly to fill his cup. "Did the pump break?" I asked. "Pump?" he replied. See https://medium.com/@sachinrekhi/don-normans-principles-of-interaction-design-51025a2c0f33

ASSISTIVE TECHNOLOGY & ERGONOMICS FOR ALL OF US

Assistive technology (AT) can be broadly defined across many contexts. In the largest sense, assistive technologies include any tool or device used by anyone to accomplish a purpose. In fact, from a human factors point of view, a *system* is defined as just that: a human using a device to accomplish a purpose (Sanders & McCormick, 1993). Conceivably, then, a human using any tool or device or machine, from a cup to a pencil to a car to a calculator to a lawn mower to a supercomputer, could be using technologies that assist in accomplishing some practical task or purpose. In this broadest sense, many of the implements or products, including tools, devices, apparatus, and machines, that we rely on each day for the activities of play, work, school, and living could be called assistive technologies. These items can help all of us to perform the tasks that we must accomplish daily in a variety of settings.

More specifically, assistive technology focuses on the unique needs of people of all ages who may have a variety of special sensory, motoric, cognitive, and/or linguistic needs. The term *AT* most often translates as special, adapted devices that allow for or enhance the participation of children and adults in many of the daily activities that most of us take for granted, such as speaking, writing, listening, seeing, eating, drinking, moving around one's home or community, using the telephone or computer, opening and closing doors, turning lights on and off. These, plus other seemingly commonplace, mundane, yet highly important tasks tend to go unnoticed by those of us who experience few barriers in accomplishing them.

A more accurate definition is necessary. For our purposes in this book, we will operationally define assistive technology (AT) as follows.

Assistive means helping, supporting, and aiding in accomplishing practical functions, tasks, or purposes for persons of all ages. Individuals who use AT may include anyone. These persons may have a variety of special needs, disabilities, limitations, and/or challenges that limit their participation in life and thus require supportive functions from other humans and from special tools and devices.

Technology means reliance on simple as well as potentially highly complex tools, devices, and equipment, and on related industrial processes, which may be mechanical, electronic, electromechanical, or hydraulic in nature (or combinations of these features), as well as the strategies, methods, and techniques that the human must bring to the interaction to make tools and devices operate to accomplish a purpose.

BRIEF HISTORY OF ASSISTIVE TECHNOLOGY

We tend to think of technology, and particularly the more specialized field of assistive technology, as a recent human development. This perception is inaccurate. People in many cultures throughout history (and prehistory) have been brilliantly creative in adapting, developing, and using special tools and devices to help others in their societies with special needs. In its broadest definitions, and even in its strictest sense, assistive technology has been used for thousands of years in numerous cultures everywhere on earth. The phrase *assistive technology* may have been recently coined, but the concept of assistive technology seems to be as old as human ability to innovate and invent (Trease, 1985).

Around 600 B.C., Aesop, the ancient Greek storyteller, taught "Necessity is the mother of invention." Before and since Aesop's time, necessity has often evoked the unique type of problem solving that we humans engage in when helping to meet, through design of special tools, the special challenges posed by disability and limitations of ourselves, our families, and our societal members.

James and Thorpe (1994) detail numerous ingenious assisting and enabling technological innovations developed by a variety of ancient peoples throughout the world. In their comprehensive work *Ancient Inventions*, these authors relate detailed accounts and illustrations of some of the assistive tools and procedures used by ancient peoples. For example, their research describes the finding of partial dentures, including metal bridgework, that were suitable for use in eating. These dentures—an assistive dental technology— were made by Etruscan craftsmen in the sixth or seventh century B.C. According to James and Thorpe, Pliny, the author and historian, mentions in his work *Natural History* that Sergius Silus, a veteran of the Second Punic War (218-201 B.C.) against Carthage, had an "iron" hand made to replace the hand he lost in battle. Other examples include their account of an artificial leg made by the Romans (circa 300 B.C.) of wood and metal parts that was found recently near the Capua area of Italy. They also tell of drinking tubes (in effect, straws), apparently invented for more convenient sipping of beverages by the early Sumerians, depicted in art from the Bahrain area of the Arabian Gulf dating from about 2000 B.C. Another seemingly modern assistive technology, glass windows, allowing the entry of light into houses for reading and work, was found in the ruins of Pompeii dating from around 60 B.C.

James and Thorpe (1994) also describe several assistive methods or strategies that were low- or no-tech in nature, but that anticipated the development of more complex technological devices later in history. Among the more notable of these was the "shouting telegraph" used by the Gauls in ancient France. Julius Caesar described this strategy for communication in some of his writings from around 52 B.C. The shouting telegraph was a highly effective, culturally embedded strategy used to broadcast news throughout the rural lands of ancient France by the rural people and farmers. Caesar described how the shouting telegraph method was used to warn the population of ancient France of the approach of Roman troops, and that it was apparently fast and highly effective in spreading the news. The shouting telegraph consisted of farmers and rural people going to the edge of their individual properties, and then shouting a short message to their neighbors, who would do the same to the next property owners and so on. James and Thorpe state that messages could travel across the countryside of ancient France at a speed of about 150 miles per day, or about 12.5 miles per hour, a significantly rapid (though no-tech) method of information transmission for sharing news in ancient times.

Interest in assistive devices and technologies continued at some level in Europe and China over the next two millennia, with the earliest documented mention (from Venice, around A.D. 1300) of what is the most popular, most widely used assistive technology yet created: optical and lens technologies. Eyeglasses (Wisconsin Optometric Association, 1996; Trease, 1985; James & Thorpe, 1994). Other accounts from these and other authors regarding the Middle Ages in Europe include exploration of assistive listening devices (ear trumpets, funnels), crutches, rudimentary artificial limbs, and other helping devices, procedures, and treatments for ill or injured persons. The idea that you could *do* something for someone with disability and/or illness began to spread and gain increased acceptance during this period in Western culture.

A true attitude shift among the learned as well as the general populace continued in medieval Europe. This shift saw beliefs modify to entertain and then embrace the radical, even heretical notion that something tangible and useful could often be done for persons living with illness or disability (Burke, 1978 and 1985). On an increasing and wider scale, people were recognizing that a treatment procedure or a tool, device, or implement (a crutch, a cart, a lens) might be of significant practical value in helping to solve the special needs of persons with physical or sensory limitations. This practical, technology-related problem solving approach was a major departure from prevailing religious thought and societal practice. Illness and disability had long been viewed in Western and other religious cultures as part of a greater divine will and a divine plan or curse for a person, with life tests to be tolerated and borne by the person and his or her family with faith and courage. Challenging, disputing, or, more disturbingly, *compensating for* or *fixing* a divinely bestowed limitation with creative, human-designed treatments or technologies was a powerful new concept popularized in and spread from medieval Europe.

The age of global exploration, begun in the late fifteenth century, initiated a period that continues to this day of increasing discovery of features of this planet and others, natural phenomena, and technology through creative trial and error, and through the evolution of the scientific method. New mechanical, hydraulic, and electrical technologies developed rapidly in Western culture, accelerating particularly in the late 1800s. At this time, the growing fields of electrical and electronics research and development saw the invention in 1876 of Bell's amplifier device for the deaf, which subsequently became what we know as the telephone. A myriad of other innovations developed in this period have influenced consumer and assistive technologies through the present day. The advent of the mechanical computer in the 1800s, the electronic computer in the 1940s, and particularly the microprocessor-driven devices of the mid-1970s and thereafter, have made "servant" technologies available to most in the industrialized world.

The continuing story of enabling and assistive technologies focuses more than ever on refinement of digital microprocessor devices combined with mechanical, hydraulic, and electromechanical tools and equipment. These devices and related materials have conferred substantial power on general consumers along with rehabilitation and special education professions to heal, make whole, and restore human capabilities in new ways. The gap between the full, normal-range function of our standard, endosomatic (within the body) equipment of eyes, ears, voice, speech, hands, fingers, arms and legs, and the compensatory functions assisted by external, exosomatic (outside of the body) technologies is closing. But is still huge and often ungainly. No assistive device yet works as well or is as convenient across settings as the original equipment of our own body parts and functions which AT is intended to replace. Fortunately, the gulf is narrowing, and our compensatory technologies and techniques are improving, encouraged by refinement of our tools, devices, and implementation strategies.

100 words about enabling technologies and the United States *CIVIL WAR* (1861-65)

The U.S. Civil war was among the first in history fought with enhanced firearms but without helmets or armor. Wounds from the new Minie Ball ravaged arms and legs. Fast, brutal amputations became primary battlefield treatments. Soldiers fortunate enough to return home often lived with significant disabilities as they sought to resume function. Innovations in and refinements of prosthetics increased as cultural notions of rehabilitation, restoration, and reliance on effective enabling devices grew acceptable. At great costs, technology progress can sometimes emerge from the cruel chaos of war. See https://www.fpri.org/article/2018/02/advances-in-medicine-during-wars   https://www.ncbi.nlm.nih.gov/pmc/articles/PMC4790547

COMPONENT AREAS OF ASSISTIVE TECHNOLOGY

Assistive technology as a field of practice has evolved and expanded rapidly since the significant advances in microprocessors and personal computers of the 1970s. AT has been described and defined by numerous authors from a variety of disciplines and professions, including Brandenburg and Vanderheiden (1986), Glennen and DeCoste (1996), Lloyd, Fuller, and Arvidson (1997), Lubinski and Higginbotham (1996), Vanderheiden and Lloyd (1986), and Zangari, Lloyd, and Vicker (1994), among others. The specific general component areas of assistive technology have been delineated by several authors, including Church and Glennen (1992). A clear listing and description of the component areas of AT synthesized from their categorizations, as well as those of several other authors, is instructive and useful; a combined listing and description will be elaborated here. A working, composite schema of assistive technologies, as adapted and expanded from Church and Glennen and from several other authori-

Designing For People | 53

ties, could be subdivided into these ten essential categories and areas of practice in assistive technology:

1. Augmentative and alternative communication (AAC)
2. Adapted computer access
3. Devices to assist listening and seeing
4. Environmental control
5. Adapted play and recreation
6. Seating and positioning
7. Mobility and powered mobility
8. Prosthetics
9. Rehabilitation robotics
10. Integration of technology into home, school, community, place of employment.

Each of these areas of AT on its own can make important contributions to the quality of life for its users. Each area may also often integrate with any or all other component areas of AT. Though it may be simplistic or misleading to view each area as a separate entity, we will attempt to clarify each in more detail:

1. *Augmentative and Alternative Communication (AAC)*

Although oral speech is the primary method of communication for humans, many persons with special needs and disabilities find that their speech capabilities are not adequate to meet their daily needs for exchange of meaning between themselves and others. A helping, nonspeech method, strategy, or device may be useful in supporting, augmenting, or, less often, replacing their oral communication efforts. These strategies and devices may range from use of manual sign language and finger spelling to a low-tech pointing board with pictures, symbols, or words, all the way to extremely high-tech devices such as portable voice output communication aids (**VOCAs**) that allow the user to speak by means of computer-generated synthesized speech or by playing back the oral speech of another human speaker that is stored in digital form on a computer chip. Tablets, smart phones, and other common consumer devices can also be used.

AAC also may include use of augmentative (supporting) or alternative (in place of) methods, strategies, and devices to assist a communicator with their writing, typing, dialing, calculation, drawing, singing, or any other form of expressive communication. Creative adaptations and innovations, particularly involving the use of microprocessor-based devices, have been developing rapidly, with ease of use and affordability of devices improving almost daily.

2. Adapted Computer Access

Computers, tablets, smart phones, smart watches, and other digital devices that help us in so many ways are parts of life now and will continue to be so. Reliance on and benefit from these productivity devices are infused into most aspects of daily life, education at all levels, employment, and personal productivity. In the industrialized world, these are all powerful forces tools for inclusion, or exclusion for those cannot use of get them. The older conceptual model of a personal computer as a TV sitting on a typewriter has been useful to many but has changed. To operate this TV-with-typewriter combination at its fullest capacity, you had to be able to see the screen, and move your fingers rapidly and accurately over the 100 or more 1.0 x 1.5 cm keys. If you had visual difficulties that precluded you from seeing the keyboard and/or monitor screen, or if you had motoric difficulties interfering with your ability to press keys on a standard computer keyboard, you were locked out of the personal computer. You were excluded from learning and using one of the essential, expected tools for personal, educational, and vocational productivity, communication, and recreation.

Adapted computer access deals with making the computer more usable by persons who may have challenges and limitations that require some nonstandard, creative adaptations for entering and/or reading out information from the digital device. Adapted, alternative methods of entry include *Modified Direct Selection* (via mouth stick, head stick, splinted hand, etc.), *Scanning* (row-column, linear, circular, and other ways of controlling a sequentially stepping selection cursor in an organized information matrix via one or two switches), and *Coding* (entering sequenced pulses from special switches to operate the device.)

Designing For People | 55

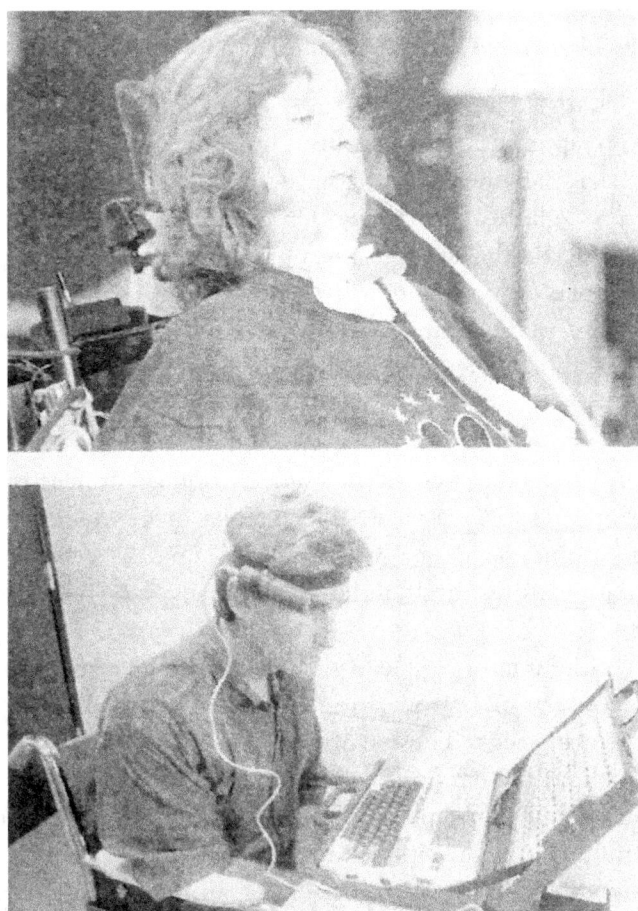

Creative solutions can encourage independent device access.

Considerable work has been done over the past fifty years to enhance the rate, lower the exertion and fatigue factors, and to increase the efficiency of adapted computer access methods. Much more research and development of products and methods are needed to make the adapted-access computer user fully competitive with other computer users across all tasks.

3. Assistive Listening and Seeing Devices

Technologies to enhance hearing and vision are some of the oldest, most proven approaches to assisting and compensating for reduced human sensory abilities. As we have mentioned, corrective optical lenses such are glasses and contact lenses, are among the most prevalent assistive technologies ever invented. From a simple magnifying glass to complex bifocal and trifocal lenses in use today, to contact lenses with exceptional capabilities in a small piece of cleverly structured plastic, optical technologies have become assumed assistive technologies, essentially invisible to the populace, and often considered fashion statements. For persons who have more extreme vision impairments, print, text, and screen magnifiers (optical or electronic) that enlarge print on paper or on a tablet, phone or computer screen can bestow visual independence where it did not exist before. Visual enhancement features and auditory readback are built into most devices now.

Assistive listening devices can include simple acoustic funnels such as the ear trumpets of several centuries ago, as well as the most modem electronic hearing aids. These can include hearing aids that are body mounted, behind the ear, in the ear, or in the canal. Assisted listening devices can also include captioned screen readers, amplified telephones, auditory trainers, FM and inductance transmission systems, and other devices providing electronic and acoustic amplification of sound, or visual alternatives. In general, assistive seeing technologies magnify and clarify images. Assistive listening technologies amplify intensity of sound only. Clarifying of sound has improved but is not perfected. The two technologies seem analogous in function, but they are not. Sharp, important differences exist in their attempt to restore or compensate for sight hearing.

4. Environmental Control

Life in a technology filled society involves moment-by-moment reliance on our abilities to control our immediate surroundings by turning lights on and off, operating appliances, adjusting the heating and air conditioning, controlling the radio, television and entertainment devices, plus myriad other operations of important switches and controls we take for granted. Disabilities, challenges, and limitations of various types can deprive many persons of these basic control skills. Adapted access to environmental control through smart phones, tablets or dedicated special devices known as environmental control units (ECUs) can confer or restore at some of the choices most of us enjoy in our personal environments.

Environmental control can be enhanced via smart phone and tablet apps, and via dedicated EC devices employing a variety of selection methods. Transmission of ultrasonic, infrared, or radio signals activate receiver units built into or connected with diverse electrical devices in a user's living, learning, play and work environments.

5. *Adapted Play and Recreation*

Play, recreation, and leisure activities for adults and children with normal- range capabilities generally involve moving, sensing, and controlling two- and three-dimensional shapes, objects, and toys, as well as interacting with people. Fun and learning occur through play and recreation. If indeed play is the work of children (Musselwhite & St. Louis, 1988), then exclusion from play and recreation certainly limits the important work that children (and, we would argue, teens and adults also) should be doing to develop all their abilities.

Difficulties with moving, manipulating, touching, seeing, hearing, tasting, and otherwise becoming involved with toys and play activities exclude a disabled child or adult from many important learning experiences that can have a critical impact on cognitive, linguistic, motoric, sensory, and social development and achievement. Battery-operated and other toys that can be modified for operation from a single switch or control, or specially designed software that allows on-screen learning through adapted access to the computer are examples of adapted strategies to allow play for those who may not otherwise be able to participate in these important, enjoyable learning experiences.

6. Seating and Positioning

Seating and positioning underlie all else we do with tools and technology. To use AAC, ECU devices, tablets, phones, adapted computers or toys, particularly over any length of time, a child or adult must be able to achieve the postural security necessary for performing related control activities successfully. Students and clients must be able to see, hear, move, or react to their interactions with devices and with human communication or play partners as well as possible in all settings and in all contexts. This optimization generally includes attention to how these special-needs learners are seated or positioned on the floor, or at a desk or table in a chair or wheelchair. It also includes how their posture is oriented and secured for device use and human interaction in any other position they may occupy during the day, such as in the bathroom, in bed, in the pool, and on the couch. All the other places and positions from which any of us may play, learn, work, and interact may be included in seating and positioning considerations.

Positioning and seating considerations are crucial to AT success. A child or adult will not learn, use, or try to become competent with any component area of AT if they are caused continuing discomfort in its use, or if they cannot see, reach, hear, touch, or otherwise interact with a helping technology from where they sit, stand, or lie.

Designing For People | 59

Adaptations to toys and other devices allow children to play, learn, and develop skills with simple assistive technologies. Whether new and high-tech, or older and low-tech, cleverly adapted toys and devices can foster learning across physical, cognitive, linguistic, and social domains, enhancing independence and confidence.

7. Mobility and Powered Mobility

Mobility means moving within and around our immediate surroundings, and being able to move among external environments, traveling as we wish. Technological adaptations can help someone move from side to side in bed, can allow people to be more mobile within their room or home, and can extend to motor-driven and human-powered devices that allow enhanced ability to move around home and neighborhood, plus travel to school, job, and recreation. Lifts, stairway guides, wheelchairs, scooters, and powered wheelchairs are among the AT items that can allow for increased mobility at home and away from home.

Adapted automobiles, vans, buses, and other transportation aids can create access to travel, without which opportunities for education, employment, and recreation could be severely restricted. Although numerous barriers still exist to mobility in most societies, the personal technology to provide for enhanced movement capabilities at home and elsewhere is available and highly useful.

Simple devices can enhance mobility for recreation, work, and socialization. Low-tech tools can serve well, with high-tech mobility solutions also offering independence for many users. Combined with assistive hiking poles, this hiker's assistive gear of protective eyewear, plus headwear, garments and footwear are technologies ready to open a world of adventure, work, and fun throughout changing seasons on varying terrain.

8. Prosthetics

Applications of technology to strengthen and/or replace the function of limbs and other body parts are developing rapidly. From splints to cochlear implants to myoelectric hands and arms to mechanically sophisticated ankles and feet, replacement parts for the body are becoming more widely available and better accepted. Prosthetic compensation can provide a viable, effective alternative to the natural function of many body parts for persons of all ages. Limb-deficient children and adults may often choose to use prosthetics in combination with other areas of assistive technology, such as adapted computer access, to accomplish tasks. Further refinements in the technology and in the integration of prosthetics with other areas of AT are underway, as are improvements in funding and acquisition of these devices for those who need them.

9. Rehabilitation Robotics

Just as many persons who have disabilities are assisted by specially trained helper dogs or monkeys, so may many be helped by intelligent robotic devices. These programmable aids and robodogs move and work in three dimensions, as opposed to the limited, 2-D on-screen assistance that computers provide. The tasks that rehabilitation robotics could assist with range from personal care to housework to educational, recreational, and vocational applications. This growing, exciting area of AT holds much promise. Also. it will be interesting to watch what happens with the new housekeeping and messenger robots, and the flying drones that offer views and explorations beyond what standard human eyes and ears can provide. These smart devices are setting precedents for friendly, affordable robotic help all of us will likely fins places for in our homes and schools, as well as at work and for leisure activities.

10. Integration of Assistive Technology into Home, School, and Community

General consumer technologies of communication, including telegraph, telephone, facsimile, email, texting, graphics, newspapers, magazines and more, plus personal, and public transportation on land, air, and water, and recreation of all types can allow the general public to move about and participate in the world at will. Likewise, integration of assistive technologies into all the component areas mentioned above can improve life for persons with special needs and limitations. In the aforementioned areas, and in some not mentioned or perhaps not even thought of yet, people with limitations, disabilities, and special needs are participating *more fully and at will* in increasingly wide sectors of life in many countries, because of transparent assistive and general consumer technologies incorporated into our everyday lives. Applications of human factors basics have made this happen as we become more aware of inclusion and universal design in our world societies.

Adaptations to building entryways, elevators, desks, and screen-reading speech synthesizers on computers, tablets and phones are now common. These increasingly transparent and omnipresent technological onramps to life can provide opportunities for freedom of association and choices of activities for persons with disabilities, approximating even equaling the freedoms others regularly enjoy.

Integration of diverse technologies into employment and learning environments can enhance safety and productivity. This woodworker benefits from wearing a mask and his powerful vacuum system that removes dust from his shop. His glasses and safety goggles enhance his safe, innovative creations of new wood art.

PRIMARY USERS OF ASSISTIVE TECHNOLOGY

Any of us could be said to benefit from technologies that assist our work and other tasks. In that sense, all of us are AT users. When assistive technology is defined operationally as we have here, however, we can more closely describe who the users are. In general, many AT/AAC users with special needs, limitations, and disabilities might be categorized into these four major representative groupings, as adapted from Vanderheiden and Yoder (1986) in Blackstone (1986):

1. Congenital and/or developmental special-needs conditions such as cerebral palsy (CP); severe sensory (vision or hearing) deficits; oral, verbal, or limb apraxias; cognitive and/or behavioral disabilities; and other areas.
2. Adventitiously acquired disabilities due to accidental trauma, surgery, or illness, such as amputations; spinal cord and/or brain damage; injuries to eyes, ears, hands, or brain; laryngectomy; glossectomy; and others.
3. Progressive neurological, neuromuscular, and other diseases such as amyotrophic lateral sclerosis (ALS), multiple sclerosis (MS), muscular dystrophy (MD), Parkinson's disease, Huntington's chorea, AIDS, syphilis, parasite infections, and others.
4. Relatively temporary conditions such as Guillain-Barre syndrome; surgery, acute trauma injury and/or burns to feet, legs, arms, hands, mouth, eyes, or face; tracheostomy due to accident or surgery; partial laryngectomy and laryngeal damage/surgery; eye and/or optic nerve infections; Reye's syndrome; and others.

Each of these major component areas of AT may have types of users and underlying disorder etiologies that occur more frequently than others. So far, little organized data is available across the categories described here. These ten categories of AT, though certainly generally supported by the work of many, are largely arbitrary and perhaps not a fully useful way of grouping human beings. Some specifics about the more frequent users and user types of various assistive technologies would, nonetheless, be interesting and probably useful. Romich (1992) has stated informally that perhaps the most frequently occurring etiology for users/purchasers of Prentke-Romich Company voice output communication aids is cerebral palsy. This company has been one of the major founding forces in much of the assistive technology and augmentative and alternative communication movements worldwide for some thirty years. Romich's information is valuable especially because it is derived from one of the largest applied longitudinal and demographic bases available. Personal experience of this author and other immediate clinical sources in the United States confirm Romich's statement.

In another organized approach to determining who AT users are, Shane and Roberts (1988) described some of the most common etiologies underlying severe writing, typing, and computer access disorders. They listed four essential categories of disorders that include: general neurological (cerebral palsy), central nervous system, peripheral nervous system, and musculoskeletal system. We have adapted and expanded on this general schema, and have summarized common AT-related needs, etiologies, and the general types of AT that may be utilized with them.

TABLE 3-1. SPECIAL NEEDS & REPRESENTATIVE USES OF ASSISTIVE TECHNOLOGY

Special Needs and Representative Uses of Assistive Technology

Need	Onset	General Category	Usual Prognosis	Some AT Types Used[a]
Amputations and limb differences due to accidents, surgical treatment, and/or birth anomalies	Acquired or congenital	Musculoskeletal	Stable	2, 4, 3, 7, 8
Amyotrophic lateral sclerosis (ALS)	Acquired	Neuromuscular	Progressive	1, 2, 4, 5, 6, 7
Arthritis—juvenile rheumatoid	Congenital	Musculoskeletal	Progressive	2, 4, 7, 10
Arthritis—osteo	Acquired	Musculoskeletal	Progressive	2, 4, 7, 10
Blindness and low vision	Acquired or congenital	Sensory-peripheral or central	Progressive or stable	2, 3, 4, 5, 10
Cerebral palsy (CP)	Acquired or congenital	Neuromuscular	Stable	1, 2, 4, 6, 7, 10
Cerebral vascular accident (CVA)	Acquired	Neurological—central nervous system (CNS)	Stable	1, 2, 4, 6, 10
Cognitive delays and/or disabilities (CD)	Congenital (can be acquired)	Neurological—CNS	Stable (can be progressive)	1, 2, 4, 5
Dementia of Alzheimer's type (DAT) and others	Acquired (?)	Neurological—CNS	Progressive	1, 4, 5, 7
Fibrositis	Acquired	Musculoskeletal	Progressive	2, 4, 7, 10
Guillain-Barré syndrome (GBS)	Acquired	Neurological—peripheral nervous system (PNS)	Progressive (may improve)	1, 2, 4, 5, 7, 10
Hearing impairment/deafness	Acquired or congenital	Sensory, PNS, or CNS	Stable (can be progressive)	1, 2, 5, 7, 10
Multiple sclerosis (MS)	Acquired	Neurological—CNS	Progressive	1, 2, 4, 7, 10
Muscular dystrophy (MD)	Congenital	Musculoskeletal	Progressive	1, 2, 5, 6, 7
Myasthenia gravis (MG)	Acquired	Neurological—PNS	Progressive	2, 4, 5, 7, 10
Polio and postpolio syndrome	Acquired	Neuromuscular	Progressive	2, 4, 5, 7, 10
Presbycusis and presbyopia	Acquired (?)	Sensory-peripheral	Progressive	2, 3, 4, 5, 10
Spinal cord injury (SCI)	Acquired	Neurological—PNS	Stable	2, 4, 6, 7, 10
Traumatic brain injury (TBI)	Acquired	Neurological—PNS	Stable	2, 4, 6, 7, 10
Wernig Hoffman's disease	Acquired	Neurological—PNS	Progressive	1, 2, 4, 5, 6, 7

Key to Types of Assistive Technology: 1. AAC, 2. Adapted computer access, 3. Assisted listening and seeing, 4. Environmental control, 5. Adapted play/recreation, 6. Seating/positioning, 7. Mobility/powered mobility, 8. Prosthetics, 9. Rehabilitation robotics, 10. Technology integration/worksi access.
[a]Other types may be included.

As noted earlier, the important concept is that technologies which assist are relevant to everyone, not just to persons living with disabilities and special needs. We all benefit from, indeed depend on, a variety of technologies to help with our daily living, plus our education, work, and play, whatever our age or lifestyle. Eating and storing food, cooking, personal hygiene, communication, travel, and access to entertainment are all certainly enhanced by the now-common technologies we find in our kitchens, bathrooms, recreation rooms, and garages.

How different and how *hard* life would be for most of us in the industrialized world without these helping technologies that we take for granted. Many of these technologies are also transparent enough to be readily usable by many persons with special needs on a regular basis. Think of curb cuts, screen captions, elevators and switch-operated doors as examples that are so common, we seldom notice them anymore. As consumer technologies evolve and develop to include an ever-wider array of general consumer and special-needs users, such as we have witnessed with elevators and telephones, as just two examples, the transparency and accessibility of general technology that assists us all will be enhanced, even for those who need assistive technologies.

The best assistive technologies, we would argue, are the ones that become invisible in their use. That is, the familiar technologies we use as tools to accomplish some purpose while concentrating only on that purpose, not on the technology itself. Typing a quick text or snapping surprise photo on our mobile phones is so easy and automatic that we forget how awesome our advantage is of high-tech in our pockets. The more we can take a technology for granted, and the more that general consumers as well as those with special needs can continue to do so, the more transparent, the more effective our use of AT in all its forms becomes.

THOSE WHO WORK WITH ASSISTIVE TECHNOLOGY

As is evident from the scope of assistive technology components, this is an eclectic, interdisciplinary field, drawing on the insights and expertise of many professionals and other personnel. AT is an area of practice that seems particularly to evoke the creativity of users, family members, and other nonprofessional but highly innovative inventors, designers, and problem solvers; input and ideas from everyone are welcome in this emerging field. A listing of AT personnel (in alphabetical order) would include but not be limited to:

AT users, parents, and families
Administrators of educational and clinical programs
Audiologists and hearing device specialists
Clinical researchers

Dentists—general dentistry and other dental specialists
Educators and special educators of all types and at all levels
Engineers, including computer, electrical, materials, mechanical, and others
Nurse practitioners and other nursing professionals
Ophthalmologists, optometrists, and opticians
Occupational therapists
Physiatrists
Physical therapists
Physicians—general practice and specialists from various disciplines
Prosthetists
Product and device researchers, developers, designers, suppliers, and vendors
Rehabilitation engineers and technologists
Social workers
Speech-language pathologists
Vocational rehabilitation counselors and evaluators.
And more....

We might also include in this listing lawmakers and public policy officials at local, state, and national levels, plus insurance company officials and others who make, interpret, and enforce the multitude of regulations that govern acquisition and use of AT in the United States and in other countries throughout the world. These personnel may be behind the scenes, but their influence is felt deeply in AT practice. Undoubtedly there are other professionals and personnel who also should be included in this listing, and we acknowledge their importance and diverse contributions to AT practice. To clarify the roles of many of these types of personnel, let us look briefly at the nature and scope of their respective roles in AT.

1. *AT users, plus their parents, families,* and other interested persons in the users' lives are often the critical creative forces behind practical problem solving, device acquisition, and funding. They are particularly crucial in the day-to-day maintenance

and implementation of AT in the user's real-world life. AT users and the other people involved in their lives can generate many innovative solutions, adaptations, and enhancements with special tools and technologies, which can often be refined and expanded on by the AT professionals with whom they consult.
2. *Administrators of educational and clinical programs* assure the continuity and quality of AT services delivered to children and adults. Administrative personnel in both settings supervise other professionals, monitor and advocate for funding, and provide leadership in other areas of public policy adherence for provision of quality AT services at their practice setting.
3. *Audiologists and hearing device specialists* are the experts in evaluation of consumers for the acquisition, correct fitting, purchase, and proper, safe use of assistive listening devices of all types. Audiologists also may often have expertise in non-speech communication methods such as signing, finger spelling, and use of specialized telecommunications devices for the deaf.
4. *Clinical researchers* from a variety of fields and professional backgrounds provide the essential foundation of information on which the rest of clinical practice and device development is based. Clinical researchers continue to offer further insights into human development, behavior, and disorders—and, specifically in the AT fields, how humans can more successfully interact with and use all types of assistive devices. Researchers also develop and test new methods, strategies, and devices.
5. *Dentists in general dentistry and other specialties* can assist AT users in development and fitting of special mouth-controlled devices such as tongue- activated switches, bite switches, mouth sticks, and other items that may help users operate computers and other assistive technologies. The mouth is an important control site for some AT users, and dental professionals are important in helping users maintain a

healthy mouth while using it as a central, long-term control site for device operation.
6. *Educators and special educators of all types and at all levels* are generally the professionals who carry the main responsibility for making AT work in their students' quests for attainment of literacy. Teachers, classroom aides, and other school personnel almost always have the most frequent and longest professional contact with child AT users. Educators are probably the most important people in assuring that AT is successfully integrated into the child's daily life of learning, playing, and socializing. They are the professionals who make AT work in the daily young lives of many users. Adult educators occupy a role of similar importance with older consumers, and can provide continuing education for the professionals, families, and others who work directly with AT users.
7. *Engineers, including computer, electrical/electronic, human factors, materials, mechanical, and others,* are essential in creating, designing, and perfecting the assistive devices and control interfaces that AT users rely on. The research and applied work of engineers are crucial in development and refinement of safe, effective assistive technologies across all AT component areas.
8. *Nurse practitioners and other nursing professionals,* often in home-visit roles, can provide important monitoring of how a user is benefiting—or not— from inclusion of assistive technologies in his or her daily life. Nursing professionals can offer important information, training, and consultation to AT users and their families regarding essential issues of how the human patient and the assistive devices are combining to become systems. These issues may include refinements in seating and positioning for mobility, cleaning and maintenance of special shunts and valves used for speech in laryngectomy or tracheostomy cases, and other aspects of monitoring how the patient's body itself is interacting with the technologies used.
9. *Ophthalmologists, optometrists, and opticians* can test, evaluate,

treat, and often significantly help improve a user's visual capabilities. Vision is critical in the successful operation of many assistive technologies (perhaps particularly so in AAC and powered mobility), and greater inclusion in the AT service-delivery cadre of vision professionals is a pressing need.

10. *Occupational therapists* (OTs) have a central role in assessing, improving, and helping to integrate motoric and sensory functions of the AT users' hands and other potential control sites on the user's body. OTs also can evaluate for and help assure achievement of the AT user's postural security through adapted seating and positioning for AT access and use by providing leadership in addressing seating and positioning issues for their clients over a wide variety of daily activities.

11. *Physiatrists* are physicians who specialize in rehabilitation medicine. Their expertise lies in the treatment and management of physical disabilities with their patients. They can evaluate for, recommend, and prescribe specific assistive technologies and devices to benefit their patients across a variety of AT areas. Physiatrists collaborate with a variety of related rehabilitation professionals.

12. *Physical therapists* (PTs) help AT users develop muscle strength and control and achieve postural security for successful long-term AT use. PTs practice in the area of physical medicine, which can include close consultation with other rehabilitation professionals in the AT team of experts.

13. *Physicians,* including those in general practice as well as specialists from the disciplines of allopathy, osteopathy, chiropractic, and other areas, often serve as the gatekeepers for access to AT when it is funded by medical insurance and other health care resources. Physicians can evaluate patients for rehabilitation potential, recommend and prescribe assistive technologies, and often rely on highly educated professionals such at OTs, PTs, and SLPs for details about a patient's suitability for acquisition and use of AT.

14. *Prosthetists* evaluate for, design, make, and fit custom

prosthetic devices to replace the function of limbs, hands, feet, and other body parts lost through trauma or disease. The work of prosthetists is of considerable functional value to the patient in helping to restore limb function for mobility, environmental control, and other essential component areas of AT. Their expertise can also have cosmetic benefits for a consumer, including enhanced self-esteem and confidence.

15. *Product and device researchers, developers, designers, manufacturers, suppliers, and vendors* are the personnel who discover, make, and sell the AT devices on which we and our clients rely. Across all areas of AT, reputable, reliable, responsive product developers, manufacturers, and sellers can be some of the most important links in the service delivery chain. In fact, they are often the last link in the chain in that they deliver the assistive technology to consumers and may help them to get the most from it after purchase. Their work assures that the right product configured in the correct way, per recommendations of all the other professionals mentioned here, gets to the user—and that it gets repaired or modified as needed to ensure ongoing successful use of the assistive devices.

16. *Rehabilitation engineers and technologists* incorporate their knowledge of human abilities and disabilities with their knowledge of mechanical, electronic, and hydraulic systems to help meet the special needs and limitations of their clients. The research and clinical practice of rehabilitation engineers are focused specifically on developing and perfecting technologies across all areas of AT, and on ensuring that humans interact successfully with these special devices.

17. *Social workers* may formally or informally monitor a client's practical use and progress with AT in home, school, and work settings, as well as a family's or employer's response to inclusion of AT into the home or workplace. Social workers can be excellent resources in the search for and acquisition of funding sources for assistive technologies across all

component areas, and in linking potential users and families with professionals of many types in the AT field.
18. *Speech-language pathologists* (SLPs) play a primary role in the selection and teaching of symbol vocabulary systems and enhanced interaction strategies used in augmentative and alternative communication. Their role may include evaluating and consulting in a variety of areas dealing with unaided or aided augmented and alternative methods of speaking, writing, and using computers. The work of the SLP focuses particularly on enhancing expressive literacy using many assistive communication technologies to improve users' abilities to speak, listen, read, write, and think using language. SLPs also work in adapted computer access.
19. *Vocational rehabilitation counselors and evaluators* work primarily with adult clients who, because of illness or injury, have not been able to work or are attempting to return to work. Vocational rehabilitation professionals can assist clients and employers in the successful acquisition and implementation of AT in the workplace to help clients get or keep productive employment.

The roles of all those involved with assistive technology are many and varied, as we have seen. There are undoubtedly others involved who have been omitted here, but whose influence in the AT team is still important. Roles and personnel described here also will likely change over the years as new technologies are developed and as professions evolve their contributions to AT practice.

WHO ARE USERS OF ASSISTIVE TECHNOLOGY?

The central question is: Who uses and benefits from human factors in assistive technologies? The answer is that, in addition to the user who can directly benefit from AT by enhanced participation and independence in life, *all of us* derive benefits. We all do better when we all do better, irrespective of abilities.

Of the millions of world residents who are severely disabled, a high percentage can likely benefit personally from the use of various assistive technologies. Each person who successfully uses any of the component areas of AT becomes a bit more involved in his or her own care and daily life. Additionally, many users can extend their growing independence into their lives at home, school, and in the community, as well as in employment settings. Not only can AT foster greater life satisfaction for the user, it also can help foster increased independence and productivity for many. Personal care, communication, and mobility tasks, for instance, that otherwise would have to be performed largely by a human assistant, often can be performed by assistive devices and technologies via independent operation of the AT user. This allows greater inclusion in the flow of life for many and can lead to enhanced independent living capabilities and gainful employment.

In most cultures, the more independent individuals can become, the less need they have of human assistance and the more likely they are to be able to earn a living and support themselves. Cultural, ethnic, and religious differences may affect this potential role of AT in users lives. Professionals must be alert to and respectful of these possible differences. Many cultures view the supportive and caring role of the family as primary and may actively resist or passively thwart efforts to make a family member who has disabilities more independent from them. The role of the traditional male or female head of the family, the permission he or she must grant for AT use, the overall technological literacy of the consumer and family, and other cultural or religious values must be considered thoroughly before AT is introduced into lives. We do not force AT on anyone.

Although complete independence and employability are realistic goals for many individuals, and although these life changes may still be dreams of the future for other potential AT users, some have had their lives influenced dramatically. These persons are examples of the promise of assistive technology. The cases of Lynn and Ron (composites of actual cases) will help illustrate.

CASE EXAMPLES

Lynn

Lynn was a 51-year-old patient in the late stages of amyotrophic lateral sclerosis (ALS) when assistive communication technology first came into her life. Lynn was living at a care center in a rural area of a state in the United States and was receiving primarily essential custodial and medical care for several years as her disease worsened. During the past two years or so, she had lost all ability to communicate, or so most people around her thought. She tolerated the daily routine of things done to and for her with no direct participation on her part. As is so common and tragic with ALS, she was still alert and capable of communication, but no one thought to tap this potential nor knew how to do it in her setting.

A visiting speech-language pathologist met Lynn one day and inquired about her case. The SLP realized through informal assessment that Lynn probably still had considerable communicative ability and began to explore ways to help reestablish Lynn as a communicating human being. The SLP, finding that Lynn had reliable up and down movement of her lower lip and could still blink her eyes, immediately established a reliable eye-blink code to enhance Lynn's yes/no responses. The SLP also realized that this simple binary movement capability of Lynn's could be harnessed, translated, and amplified further via assistive technology. By using a visual chart to teach Lynn the binary movements for Morse code, the SLP soon had Lynn entering Morse code commands to an adapted computer with her lip movements. The computer thought she was typing from the keyboard, and in an hour or two of practice, Lynn learned to enter Morse code to the computer. With her reliable up-down lip movements on an adapted switch, she was now able to speak using the speech synthesizer on the computer. She also could write on the computer by emulating keyboard functions with her encoded switch entries. Lynn now writes letters to her family members, composes prayers and meditations for the weekly prayer services in her nursing home, and has written her will, all through the power of assistive technology and a creative clinician who was willing to experiment with tech.

Ron

Ron was an All-American athlete at a Midwestern university. As a junior at this school, he was a business major who also enjoyed swimming and diving as sports, and painting portraits and landscapes as his primary hobby. During the summer between his junior and senior years at the university, Ron was swimming at a pool and diving off the high diving board with some of his friends. On one dive, he slipped as he took off and landed partly on the pool deck as he fell. He broke the second cervical vertebra (C-2) in the fall, and his spinal cord was permanently damaged at the C-2 level. Ron's life was saved by skilled personnel at the pool and in the hospital, but his dreams of finishing his university degree and continuing his art were dashed forever with his fall. Or so he and his family thought.

After Ron became medically stable, he was taken to a large rehabilitation hospital in the western United States that is well known for its aggressive, creative applications of AT in the rehabilitation of spinal cord injuries. The comprehensive team of rehabilitation professionals worked with Ron and his family to innovate ideas for his rehabilitation and to apply their creative problem-solving skills using AT to his numerous special needs.

Ron retained some head and neck movement after his fall and could still speak even though his breathing now was regulated by an external ventilation device connected to the tracheostomy tube inserted into his airway. The team first got Ron a simple off-the-shelf landline speakerphone from a local discount store and fitted him with a mouth stick. With the phone nearby, Ron could receive and make phone calls, thus reestablishing him as a communicating person with his relatives and other friends back home. Next, the OT, PT, and rehabilitation engineering staff evaluated Ron for a powered reclining wheelchair, controlled with head movements from specially placed switches. Ron learned to use his powered chair well and soon was able to move around the hospital and grounds with increased independence. This rehabilitation team also configured a computer workstation for Ron so that he could use a speech-input system for computer control, allowing him to write, type, and print from his desktop computer.

They also positioned a large trackball-adapted input device within reach of his mouth stick so that he could operate the computer with his mouth stick and continue to draw and paint on screen using graphics software. After six months of recovery and special training with his new AT systems, Ron returned to his home community. He reentered university and completed his business degree, doing all of his own studying, note taking, writing, and test taking. With his increased sitting time, as he called it, he focused more on drawing and painting using his mouth stick/trackball-operated computer system. His drawings and paintings have become more refined, and he has developed into a featured local artist in his area, making sales of his work at area art shows.

---

100 words about *DEEP WORK*

Cal Newport describes Deep and Shallow Work. Deep Work, harder to do, is accomplished in periods of intense, uninterrupted concentration. Shallow work, easier to accomplish, happens amid distractions and interruptions. Newport contends much of our workdays are spent in Shallow Work, and explains four key rules for Deep Work: Work Deeply, Embrace Boredom, Quit Social Media, and Drain the Shallows. Deep Work for shorter, intense periods, is stated as more productive than longer hours of Shallow Work. Careful attention to workplace safety, human factors and ergonomics are likely to reduce errors, distractions, and interruptions. See https://www.calnewport.com/books/deep-work

---

SUMMARY

The Old English word *healan* is the root word of our words *whole* and *heal*, per the *Online Etymology Dictionary* (2021). General and specialized assistive technologies can indeed enhance our abilities as professionals even general consumers to heal, restore, and help make our clients, students, and patients whole again. Indeed, the impact of tech can sometimes be dramatic, even inspiring. In all cases, however, it is often easier to be more attracted to, impressed by, and focused on the gee whiz of the high-tech tools and devices than on the human user. We must remember to ADD HUMAN. Assistive and consumer technologies, no matter how high-cost or complex, have value only in their successful interaction with real people, often in challenging circumstances. Careful consideration of critical human factors can influence whether use of tech is successful or not. The rest of this book examines many essential human factors in assistive technology.

REVIEW QUESTIONS

1. List several tools or devices that assist you in your daily life. Is each one transparent, translucent, or opaque in its use? To you? To others?
2. What about its makes each so initially? After a period of use and practice?
3. Of these assistive devices, which do you use most? Least? Why? What makes you prefer some designs over others?
4. Describe a tool or device that was opaque to you—but not to someone else—the first time you tried to use it. How can you explain the difference?
5. You are designing a new device to be used by people with variety of needs and abilities. What is your device? What will it do? Can you assure its transparency for all users? How?
6. Do you see a role for learning and training in AT success? Can training and practice help make an opaque device translucent or transparent in its use? How? For whom?
7. What assistive devices or technologies do you know of that have been created by people of different cultures? Can you name describe at three examples?
8. We discussed ten general categories of assistive technologies. Can you give one example for each area of a specific device or tool that you have used or have observed someone using? Where was it used and how?

9. Who are the users of AT that you have met? What types of disabilities or challenges did they have? What component categories of AT did they rely on?
10. Are there other component areas of AT? Name and describe them.
11. With which AT professionals have you interacted? What were their roles? What do you believe their roles should be in AT?
12. What other professions and resources should be added to this listing? Why?

4

HUMAN FACTORS GOALS

If all my possessions were taken from me with one exception, I would choose to keep the power of communication, for by it I would soon regain all the rest.
Daniel Webster

All of us seek efficacy and safety in our devices, and the reasons consumers and customers, plus our students, clients, and patients use technologies are many and varied. Goals for implementation of tech use can extend across all the component areas we have described and can vary depending on the needs and nature of individual students and clients, as well as on the professional clinicians, teachers, and others who are working with them. It is unlikely that we could delineate all possible goal areas for AT across all possible clients and AT practitioners. Nonetheless, identification of some concrete, specific areas for attainment via the power of assistive technology that may share some universal recognition across fields and clients is valuable. These goals may not generate universal agreement across fields, but they can serve as springboards for further discussion and refinement of our collective purpose in AT practice.

One attempt to define specific goals in the AAC component area of AT was the listing presented by Shane (1986) in Blackstone (1986).

Shane developed nine goal areas specific to augmentative and alternative communication intervention that are large enough in scope to cover most subareas of practice and achievement in the AAC component area of AT. As adapted from Shane, these intervention goals for AAC include:

1. Equalize the gap between comprehension and production.
2. Promote greater participation in the school and other settings.
3. Enhance vocational opportunities.
4. Promote interpersonal and social interaction.
5. Reduce frustrations associated with communicative failure.
6. Enhance language comprehension.
7. Facilitate speech development.
8. Serve as an organizer for language.
9. Enhance speech intelligibility.

Although Shane was referring specifically to AAC intervention with these goals, most of them could also be extended to other component areas of technology use. A review of the AT component areas listed earlier, and comparison of the goals defined by Shane to those areas, indicates the universality of Shane's statements across many, perhaps all, areas of AT. Component areas of AT described in Chapter 2, although discrete on the printed page, intermesh considerably in our real-life work with students and clients.

Just how one distinctly separates AAC from adapted computer access, or adapted play from environmental control, or assisted seeing from integration of AT into the workplace in the daily lives of actual students, clients, and patients, is more of an academic than a practical pursuit. All components of AT can be closely interlinked when they are being employed with real people in real settings. Drawing sharp distinctions between and among them may not be as valuable as finding what common ground is shared across all component areas and related general goals. Goals pertinent to one area of AT practice and client use will likely be useful as platforms from which to view the other component areas of AT.

Similarly, Silverman (1995) also described goals for AAC practice that can give us insight into more global applications across all AT component areas. Some of the more generalizable of these goals for AT intervention, as adapted from Silverman, may include helping or teaching clients to do the following:

1. Make the gestures necessary for activating a particular switch or control.
2. Transmit messages more rapidly or receive messages more reliably.
3. Get help by telephone (or other technological means) in an emergency.
4. Use a computer to accomplish real work in school and employment settings.
5. Increase their desire and intent to communicate and interact.
6. Increase intelligibility of speech and written messages for a variety of purposes important in their lives.
7. Work with trained partners who interact more successfully with the AT user and with the AT devices being used.

Once again, it is evident that many of these goals have application across many AT component areas we have described earlier. In addition to these AT goal statements described by Shane (1986), Silverman (1995), and others, additional important purposes include the following extensions beyond the usual types of goals.

Assistive technology should also help the user and those around them to:

1. Enhance personal safety and well-being in meeting urgent needs and emergencies at home, at school, in the workplace, and in the community.
2. Enhance the quality of life by allowing exposure to, contact with, and interaction among greater numbers of settings, experiences, and people.
3. Enhance the duration of life by involvement in a higher quality of experiences and by increasing involvement in one's own health care via improved access to and

communication/interaction with health care providers and the health care and rehabilitation system.
4. Enhance the quality and duration of life by allowing more mentally and physically healthful control of environmental factors such as lighting, heating/cooling, humidity, hydration, and personal care abilities.
5. Have greater access to and fuller participation in the growing network of global telecommunications essential to competitiveness in school and employment, i.e. full use of phone, texting, voice mail, e-mail, internet, social media, and other networking.
6. Enhance access to and participation in local as well as global air, land, and sea travel essential for occupational, educational, and recreational pursuits as desired by an individual with special needs.

We submit these broader goals because we believe that the ultimate purpose of AT is to open the world—all aspects of it—to those of us with limitations, challenges, and disabilities. Our participation in all the comings and goings of life in our society should increasingly become identical to that of anyone else.

Large goals? Yes, but important to the full inclusion of everyone into the main currents of life in industrialized and other societies. How many other barriers in daily living, transportation, education, recreation, and movement in and control of a variety of factors in our personal environments may be eliminated by creative applications of assistive technologies?

---

100 words from Goethe about *WORK AND EFFORT*

"Lose this day loitering...'twill be the same story/
  Tomorrow--and the next more dilatory;
  Each indecision brings its own delays,/
  And days are lost lamenting o'er lost days.
  Are you in earnest? Seize this very minute--/

Boldness has genius, power and magic in it.
Only engage, and then the mind grows heated--/
Begin it, and then the work will be completed!"

From *Faust*, Johann Wolfgang von Goethe, 1749-1832. John Anster, translator.

Adapted from W. Dyer, 1998. *Wisdom of the Ages.*

HUMAN FACTORS AND INTENTS OF ASSISTIVE TECHNOLOGY

As described previously, human factors and ergonomics across all of the ten component areas of AT are concerned with how humans who have special needs, challenges, and disabilities interact with and are influenced by the assistive technologies, tools, and devices they use. Human factors in AT are concerned with matching features of the devices employed to help meet users' needs, capabilities, and limitations; they are focused on increasing user effectiveness.

efficiency, convenience, comfort, safety, satisfaction, and performance. Human factors are similarly focused on reducing user exertion, stress, danger, failure, and rejection of the system of devices they are attempting to use.

The best, most complex, most expensive assistive technologies are destined for failure if human factors are ignored in design and intervention. Whether we are planning to implement an assistive seeing technology, or powered mobility, or an augmentative communication device, or any other assistive technology, such seemingly basic aspects of these as how they appear, how many steps are needed to operate them, and how well the device is designed to prevent our errors in using it can all combine to produce success in using the device ... or failure with it, if we ignore some of these apparently mundane human factors. As these factors can conspire, in their failure or absence, to engender learned and taught helplessness, safety concerns, and just plain failure of our students, clients, or patients to accept and use the technology we have prescribed for them, they become critical to consider.

These human factors are likewise dangerous to ignore. Failure of an intended AT system (the user combined with a device to accomplish an intended purpose) means failure of our learned diagnostic and prescriptive efforts, and of our intended intervention. Failure of our intended, carefully selected AT device to become a true system with our user means we have spent precious client and family time and effort in pursuing an illusion of success; we have spent valuable professional time in pursuit of creating potential assistance from a technology whose full use did not materialize. Perhaps most seriously, we have set in motion a sequence of failed tool use, device rejection, and learned/taught helplessness that can militate against our clients ever attaining the success with AT that we had sought. This seems like a heavy burden to carry for the practitioner who is only trying to assist, and who is trying to help the person with special needs to live a fuller life.

We must recognize that when we expect persons who have special needs to take on a new tool or a new device, and to become skilled with it, even amid their other challenges and limitations, we thus simultaneously expect these individuals to do several potentially difficult things:

1. Become skilled with behaviors and devices that we ourselves are probably not currently using/doing. (So how can we understand the difficulties these expectations place on them?)
2. Add to their own already existing "handicap" by taking on the mastery of some new device with which they are similarly not skilled and with which they will experience at least some initial fear and failure, just as they so often do in other aspects of their lives (per Musselwhite and St. Louis, 1988).

Hippocrates stated, "First, do no harm." In our endeavors to select, teach, and implement tech with patients, clients, and students, this must also be our prime directive. Attention to underlying human factors can help us to do no harm and ultimately do much of value for those who put their trust in our technology judgments.

APPLICATIONS OF BAKER'S BASIC ERGONOMIC EQUATION: BBEE

As described before, a useful framework for understanding the expectations we begin to place on someone who we wish to become a successful user of assistive technology has been described by Bruce Baker (1986). Baker's equation regarding motivation and load factors in AT is a visual portrayal of major factors involved in understanding and producing user success. It allows us to see the relationship of major human factor areas in the pursuit of AT success, and to understand better the individual and collective roles of these factors when intervention with assistive technology fails.

Once again, as adapted and expanded from Baker (1986), this equation is:

$$\frac{\text{MOTIVATION TO USE AND SUCCEED WITH A TECHNOLOGY (M)}}{\text{PHYSICAL + COGNITIVE + SENSORY + LINGUISTIC + TIME LOADS (L)}} = \text{USER SUCCESS...or not (S)}$$

Key points of Baker's equation, as adapted here, are summarized as follows.

1. Technology success occurs if the upper part of the equation, user motivation (M), exceeds the sum of all load or effort factors in the lower part of the equation.
2. Failure of tech will predictably occur if the total of the load and effort factors in the lower part of the equation (P + C + L + S + T) exceeds M.
3. Across all professional fields of practice that are related to AT, our primary goals for practice in AT diagnosis, prescription, and intervention are to maximize M, while contributing from our areas of professional expertise to minimize the load and effort factors (P + C + L + S +T) individually and collectively.

Let us further examine how this might be done for each component of the equation. Case examples and discussion are offered below.

Enhancing User Motivation

Nothing succeeds like success is surely the guiding principle here. Why do most first-time, new users of helping technologies, such as telephone, car, and personal computer, begin to use and then generally continue to become skilled at using these devices? It is primarily because we have seen others using these devices, have witnessed the advantages use of these devices can confer on others—and because we wish to derive the same benefits for ourselves. Our older or more technologically experienced family members, peers, classmates, colleagues, and others typically model for us successful use of these types of technologies in their daily personal and vocational lives. From an early age, most of us witness others using tools, devices, and technologies to accomplish purposes important to them, including rapid communication, independent travel, and faster and better production of written work for school and job, among others.

In an industrialized, technological society such as ours, we become saturated with observations of others using and benefiting from ("modeling") the myriad of devices and tools that surround us in our daily lives—and the benefits of using these items. In fact, we largely take for granted all of the switches and controls in our lives, the technologies they allow us to operate, and the purposes they allow us to accomplish readily. If we wish to turn on a light in a room, we flip a switch. If we wish to leave that room, we walk to the door, turn a simple knob to open it, and walk through the doorway. If we wish to speak with someone not in our immediate presence, we pick up the handset on our desk telephone or our smart phone, push a few buttons, and are generally soon conversing with whomever we wish. The power and convenience of technology use are apparent to most of us in industrialized societies from an early age; we are motivated to use the same devices, to do the same things that we observe others using so well to make their lives easier, more efficient, and more interesting.

Is this same type of modeling and growing interest in eventual use of devices true for persons who cannot access the general technologies on which we rely so much in our daily lives at home, in school, and at work? Emphatically, no. It appears that as persons with limitations and disabilities observe others using tools and devices that they themselves cannot, a chasm of us-versus-them develops. As someone watches others use things that they cannot even touch or handle or push or get into, the realization must occur that use of those devices and the benefits conferred by them are often not for the person with special needs. The slayers of motivation are subtle but thorough. Barriers to initial experience and success with even simple devices can be perceived as huge by many disabled persons, deflating their motivation ever to use or develop skill with the things that they see others using but that they themselves cannot.

Thus, the first attack on an AT user's motivation is unintentional on the part of others. It derives from those around them, across the chasm of separation caused by disability, as participating in life in ways that a person with special needs cannot. Success with adapted, probably simple, transparent technologies for potential AT users thus becomes crucial at the outset; if other technologies are beyond their abilities to interact with right now, let us get them experience with devices that they can use successfully, right from the first or second attempt. Then, let us move them up the hierarchy of successful device use so that their motivation to try out and learn the next steps is built upon their previous success in using antecedent devices.

The self-perpetuating cycle of motivation from successful device use can start small and simple. An infant can roll to one side of his crib mattress and notice that a musical crib-side toy or fan is activated—because a small pressure-operated switch is placed under that side of the mattress to control the battery-operated toy or device mounted on the crib rail. Another infant might coo or babble, and thus activate a sound-activated toy placed near her. Slightly older children may lean their heads back on a switch to make a battery-operated toy work and begin to understand the cause-effect relationship between their actions and things that they can make happen in the world. Motivation to keep on using simple technology builds. Even older children may provide adapted access to computer games or Spin Art and other toys. They can now play as the other children in their family or school do and can be reinforced by the success. Successful use of technology and devices tends to breed motivation to make more use of the technology, and so on. As we move with a user to some of the more complex and useful assistive technologies across all the AT component areas, we must gradually introduce them to initially easy, open, transparent versions of the device use. Then, in incrementally harder steps, we help them to use, succeed with, and thus develop more motivation from AT across all areas of need. These can be at their threshold of success and can supply motivation for more—and more complex—use of technology.

Who are the professionals and others who ignite this fire of AT motivation with a child? Probably the most important persons in beginning this use-success-more use-more success cycle of growing motivation are parents, family members, and teachers. These are the persons who have the most extended contact with a child and who likely have the broadest effect on a child across all areas of daily living, play, fun, socialization, and academic learning. As AT is introduced to the child to help meet her needs, parents, family members, and teachers are the persons who can help her implement the devices in the primary occupation of children: play. Musselwhite and St. Louis (1988) and Musselwhite and King-DeBaun (1996) have stated that the work of children is play, and that to truly motivate a child to use, learn, and become more skilled with any assistive technology, one major avenue must be through play. Here are some practical suggestions for helping ignite that first spark of motivation for a person to use AT, and then to fan that spark into a self-fueling fire for more use of helping tools and devices. These points and ideas are adapted and derived from Burkhart (1997), Elder and Goossens' (1994), Goossens', et al (1992), Goossens' and Crain (1992); Musselwhite and King-DeBaun (1996), and Spacey (2016 and 2017)

1. As mentioned previously, adapted battery-operated toys and devices can be highly motivating to young children, and even to older persons, including some adults. By configuring a snorting piglet or roaring dinosaur with an appropriate single switch, a user can learn some very basic cause-effect and switch operation skills that set a foundation for success with more complex technology eventually. Some excellent references for how to adapt and use simple battery-operated

toys are available in the resources mentioned earlier, as well as in materials, books, and devices available from AbleNET, Toys for Special Children, Crestwood Communication, Enabling Devices per Bell (2019), plus Amazon.com and others.

2. Simple socially interactive games such as "Simon Says" can be powerful motivators for new AAC users (Maro, 1993). They can be Simon as they experience the power of speech output from their device by speaking Simon Says directives to other children in their play group or class. The act of using a device to produce speech output that then causes others to do things (especially silly things in a creative game of Simon Says) can be powerfully motivating for a child to continue using and developing further skill with a speech-output device. Similarly, a musical chairs type of game, wherein the AT-using child controls the music from a single-switch- operated tape player or radio, can also be a highly motivating, yet basic starting place for new technology users. The success of being the leader of others and causing them to move and stop along with switch activations of the music is a pleasantly infectious low-tech activity for many children. It teaches success and resultant motivation with simple equipment. This early success on repeated uses over time can be a steppingstone for use of more complex AT devices.

3. The traditional card game "Go Fish" involves some easy-to-learn, yet strongly motivating interaction and competition: skills from which beginning users of AAC devices or adapted computer access systems can readily benefit. The phrases "Do you have a…" "Yes, I do," "No I don't," "Go Fish!" and names of the specific fish, animal, or number cards used can be quickly stored on a user's speech output device and identified with appropriate accompanying symbol representations. These phrases can also be easily programmed into a synthesized speech-output device for similar retrieval in the game. The AT user, along with his or her play and communication partners, can then participate

in a socially interactive, mildly competitive, and—for most new AT users young and older—a motivating, interesting introduction into just what the speech-output device or adapted computer can do for them. It's fun, something you can do with others to become a more skilled and more motivated AT user.

4. As children, teens, and adults attain basic literacy skills, adapted computer access via a variety of methods can allow them to participate in individual and partner games of Matching (Concentration), Hangman, Tic-Tac-Toe, and other basic spelling, reading, and math software activities that are available from a multitude of commercial suppliers. Typically, the control skills and cognitive/linguistic loads of these basic activities can allow many beginning users to develop accessing skills and motivation to eventually pursue more complex activities through success with these basic games. Similarly, adapted access to standard Nintendo, Super Nintendo, Sega, and other games can be motivating in themselves, and can generate another level of motivation in the social interaction, discussion, and even competition with peers that may derive from involvement with these games.

5. Environmental control and mobility devices often provide their own face-value motivation. A child or adult who can now control the lights or temperature in his or her room, or who can now move about more freely with a powered scooter or chair, often needs nothing more than just continued chances to use his or her devices independently. As they can oversee their own decisions on controlling their environment, and on where to go, the motivation to keep on using these skills and to perfect them continues to build.

Reducing Physical Effort

Often, a major barrier that keeps potential AT users from success with their devices is that they must simply expend too much muscle force, too much work to use the tools they wish. Something as simple as changing the location of a switch or control so that the user does not have to reach so far or push so hard in an awkward position can create major changes in the successful use of a device. Consultation with occupational therapists and physical therapists, as well as rehabilitation engineers and others with similar expertise, may help to identify ways to make a switch or control easier to operate for the AT user. As will be discussed further in Chapter 6, these items often can be the pass keys to use of assistive technologies. If they are too hard to operate, user motivation and success will diminish. Selection of the control site or sites, the part of the body used to operate a given switch or control, may affect physical effort. Switches, controls, and devices that cannot be operated by smaller, weaker control sites such as fingers, may be modified for operation by larger, stronger control sites such as feet or knees. Or if precise on-off activation is too difficult to achieve with feet, hands, or limbs, perhaps some of the work can be transferred to less obvious controls sites such as head motion or puffing or sipping on an adapted switch.

The positioning and mounting of switches, controls, screens, and other devices can help reduce the physical effort expended by an AT user, as can precise attention by skilled professionals to the user's seating and positioning. Postural security, or lack of it, in relationship to continued operation of any device can be a major factor in how much effort the user expends to operate the device. As mentioned earlier, even though the concept seems extremely basic, we all tend to stay with and be motivated to continue using devices that are easier to operate. The jam in the jar whose lid we cannot open does not get eaten. The nut and bolt that we cannot turn does not get loosened. The bike with no gears for going up hills does not get ridden as much as the bike that can be shifted, and so forth.

Here is also where our earlier principle of evaluating a device from the actual user's point of view comes into focus once again, as described by Anderson (2021), Norman (1993), Sanders and McCormick (1993) and Kantowitz and Sorkin (1983). Switches or controls or other devices that may seem easy for us to operate as teachers, parents, or clinicians, may be much physically harder to operate when viewed from the AT user's perspective of seating, positioning, or muscle force available to accomplish operational tasks. Honoring the needs, abilities, and limitations of users, and matching physical force and positioning requirements to their individual features become the primary focus in minimizing physical effort across all types of assistive technologies.

Here are some practical suggestions to reduce physical effort:

1. Make activation sites, switches, and control surfaces larger so more mechanical advantage is gained when using them with weaker control sites.
2. Adjust the sensitivity and selectivity of switches and controls to match needs of the user. See next chapters for more details.
3. Put switches and controls for AT devices within the comfortable movement range of the user, so that, regardless of control site, he or she can touch, push, flip, or otherwise activate these without having to lean, stretch, extend, or apply pressure in ways that are uncomfortable and fatiguing over repeated uses.
4. Explore easier alternative control methods. For example, using a voice- or sound-operated switch might be much easier than using a stiff light switch on the wall, or perhaps learning to use a chin-operated joystick will be much less fatiguing than trying to resist spasticity with a hand-mounted joystick control. Other creative solutions that involve these factors can be conceived for virtually any device activation activity where too much physical effort is the concern.

---

100 words about proxemics in an architectural tale of *SEATING AND POSITIONING*

In 1912 Stuttgart, Germany, composer Richard Strauss consulted with designers on the required size of the planned opera hall orchestra pit. Strauss stated it should hold about a hundred people for a symphony orchestra. Architects soon brought in 100 military band

members, having them stand at rigid attention next to each other to compute required space. Upon rehearsals for opening night, the new orchestra pit could seat only about forty players, enough for a chamber orchestra. Architects had placed soldiers tight next to each other, not anticipating the seating and space requirements of musicians with their instruments. Truth...or a fine story to make a point about proxemics? https://www.yourclassical.org/amp/episode/2021/10/25/a-strauss-tale-too-good-to-be-true

---

Reducing Cognitive Effort

As with physical effort, devices that require a lot of thought to use tend to be the ones we are less motivated to continue using. To reduce cognitive effort, such component factors as short- and long-term memory, problem solving, and sensing and discerning of control shapes, sizes, mappings, and configurations must be considered. Reducing cognitive load has much to do with the central concept of device transparency that we dealt with at length in earlier chapters. The more obvious the method and purpose of use for a device, the lower the cognitive effort that is required to operate it. The more that a user must analyze and problem-solve to determine how to use a device, the harder it becomes to use. This may be especially true for persons who may have reduced cognitive and/or sensory abilities.

For example, a familiar model of cognitive simplicity for most users is the standard consumer two-button radio. Most standard radios operate with two main controls on the front panel, one that controls on-off and volume (intensity) of sound, and another that controls station selection by moving a pointer device across a numerical dial display of some type. The user who knows these basic three operations can then problem-solve how to use most radios across brands and types, because they will all operate in roughly the same way. Cognitive load for use of this device consists essentially of "What is on? What is off?" "How do I make this louder/softer?" and "How do I select the type of sounds (i.e., the station) I want to hear?" This exemplary model of accomplishing some sophisticated technological operations with simple movement of just two controls is something to be emulated by all technologies to be used by humans—especially assistive technologies to be used by humans who have limitations and special needs. Some practical considerations to reduce cognitive load in AT include the following:

1. Make one switch or control operate just one function. If one switch or control operates more than one function of a device, make the methods of operating the switch or control to get at those functions distinctly different from each other, and provide the user with feedback when the function of the control changes (e.g., the radio on-off button that clicks at the point where it is on, then becomes the intensity control for the radio). Similarly, consider most computer keyboards: One key does one thing unless clearly marked otherwise. No standard keyboard has function keys that do more than one thing, and no number keys or others stand for more than two operations (5 and %, 7 and &, = and +).
2. Reduce the number of steps or control operations to make the device do what you want it to do; the more steps, the harder, more complex the cognitive load. The likely reason many of us do not use all our home entertainment options.
3. Make the shape of the switch or control, or the design on it, representative of its function, or at least sensorily unique so it can be more easily remembered.
4. Have controls located in places and move in ways that relate to what they will make the device do—push the joystick to the right, the chair moves to the right, and so on.
5. Provide feedback, potentially across many modalities, to indicate that a switch or control has been operated successfully.

Reducing Linguistic Effort

Linguistic and cognitive effort factors are close relatives; many concepts that pertain to one also pertain to the other. Reducing linguistic effort commonly involves making the symbols, pictures, letters, words, or other representations of meaning as easy to decipher and follow as possible for a given user. Linguistic effort is particularly evident in adapted computer access and augmentative and alternative communication. How one decides which key to touch, which item on a scanning selection menu to activate, or which selection area on an AAC device to push is affected considerably by how meaning is represented at that activation area. If a picture is used for beginning users, is it a photograph or line drawing? Is it in color or in black and white? If it is a symbolic representation of meaning, similar considerations apply, plus some others: Are the symbols part of a "system" wherein there is a uniform, predictable, evolving method used in symbolically depicting meaning as information becomes more varied and complex; or is a symbol "set" employed, wherein the symbols have no relationship to one another other than being a collection of pictorial items that represent meaning and that are all drawn in about the same way? If a tactile system of meaning is employed such as Braille, manual alphabets, or raised dot and line orthography, how large and how distinct are the movements or markings? If an alphanumeric (letters and numbers) system is used, which language are they from? Are they upper or lower case? What size are the characters and in what type style or font do they appear? Are they arranged in an alphabetical, QWERTY, Dvorak, or other type of array? If reading of words, phrases, or sentences is needed, are they paired with a symbol? How complex in structure and how long are the phrases or sentences? These and other aspects of representing meaning on an AT device are part of the linguistic effort needed to use it.

Some practical considerations for reducing linguistic effort include the following:

1. Use pictures, photos, drawings, symbols, letters, words, or sentences that are easily visible to the user and are appropriate to the user's level of development and ability to discern meaning from them.
2. Make one item represent one thing only unless multiple meanings for a symbol item are clearly demonstrated, explained, and practiced. Users who have greater linguistic

abilities, who have stronger receptive and expressive language skills, and who are partially or totally literate can interpret, process, and act on longer, more complex groupings of symbols, letters, and words; the reverse is true for those users who are less linguistically developed.
3. Make representations of meaning on an AT device as readily discernible to communication partners, skilled and unskilled, who may interact with the user. This helps ensure that the user and partner or helper can successfully interact to refine use of the device to be even more effective for the user.

Reducing Time Load

The amount of time that it takes to activate and control an assistive device can markedly affect how much load device use places on the user. With general consumer devices, the longer it takes for us to make a device do the task for which we have acquired it, the more likely we will not use it as much. More commonly, the number of steps in use plus the wait for the device to respond are often the deterrents to our further use of the device—and, conversely, items that are quick to understand and use, and that respond to our operational commands quickly, are the ones with which we are more inclined to continue working. Time is in many ways a shorthand measure of any or all these factors combined; the more effort in these areas, individually and collectively, the more time it will take to utilize a device. The more time it takes overall to use an AT device, the less responsive it is to meet the purpose for which we intended it, whether that be communication, mobility, environmental control, or another area. The less responsive the device is, the higher the time load is for the user—and the less likely he or she is to use it. Here are some practical ways to influence time load in AT.

1. Reduce delay time from switch or control activation to device response time to the minimum that is safe and

satisfactory for a given user, especially in environmental control, AAC, and mobility. Enduring many repeated, slow activations can be extremely discouraging for us all, particularly if increased rate of response is possible to configure for a given device and user.
2. Make one activation carry through several steps (macros), such as scripting and nicknames for signing on to the internet: one or two keystrokes can complete the whole sign-on sequence, rather than many keystrokes.
3. Use abbreviation expansion, semantic compaction, macros, and other combining of operations, as appropriate to a user's cognitive and linguistic abilities, to reduce physical effort and time load in operating AAC and adapted computer devices.
4. Establish and agree on with users, communication partners, and assistants the use of nontechnological (no-tech) time-saving strategies such as completion of sentences, anticipation of ideas and words, and activation of devices in the environment for the user. But use caution in this endeavor. Be sure that the user wants and consents to others stepping in to anticipate and help in finishing his or her efforts in order to reduce time load. Many AT users *do not* like or want this type of assistance, both because the assistant often misinterprets what they really wish to convey, and because they feel demeaned by it.
5. Position controls, switches and items in the user's environment, and arrange schedules of personal care and other activities so that efficient use of time is made, with least physical effort; that is, place the telephone within close proximity of the computer workstation so communication tasks can be carried out from one location without having to move among sites; or teach a routine of face or leg shaving, brushing teeth, and using the toilet before bathing or showering so that all personal cleanup can occur at once instead of over several small steps. Assorted simple, no-tech, practical methods can help reduce physical effort.

We can readily see that the interaction of these human factors in successful use of devices can be complex, with many permutations for interaction of factors. As with all things human, individual differences across and among all these factors described in Baker's equation (BBEE) can also introduce considerable variability into the clinical situation. Each learner and technology user has different needs, intents, and situations in which he or she may wish to become successful in use of AT. Couple these variables with individual differences such as type and extent of disabilities, challenges, or limitations, plus special medical, psychosocial, educational, and vocational needs, and it is apparent this is indeed a complex mix.

How we address each variable for a given client or student undoubtedly varies with the individual and with the individual professionals and practitioners who become part of that person's AT team; the potential variations here are enormous. Identifying the personnel who can address each area of this equation is a more finite question and can be more readily discussed.

Motivation to learn and then become a competent user of an assistive technology can be derived from many sources. As we have discussed already, the main source of lasting user motivation probably comes mostly from within the individual; the self-perpetuating cycle of initial "use-success- more use-more success," and so on, takes place mostly within the AT user's own mind. The reverse is likely true, too—failure experiences and resultant learned helplessness are reactions that occur, unfortunately often unobserved and unconsidered, within the user without our knowing or noticing. What persons external to the AT user might be able to help to provide expert assistance in helping to foster motivation?

Perhaps the most important persons who come to mind are first the parent or other family members and the immediate caregivers of the person seeking to become a skilled AT user. Parents, close family members, and care providers are the persons who, at least with younger AT users, have the most daily, even moment-to-moment contact with their special-needs person. The attitude expressed by parents and these other important persons can successfully model interest, enthusiasm, and an expectation of success for the user, or they can model exasperation, frustration, and rejection of AT. Both may be powerful forces in how a potential user feels about and reacts to early experiences with any type of assistive technology introduced into the daily lives of a family. Parents and caregivers who exhibit patient, optimistic enthusiasm for acquisition, use, and improvement with a variety of technologies for their child or other family member are likely to create an atmosphere at home of interest, acceptance of initial failures, fumblings and misuses, and belief in the potential for future successes with AT devices. This positive, visionary approach is likely to engender tolerance of mistakes, a freedom of learning, and initiation of that powerful rolling-snowball effect that AT success can have in perpetuating its own use by increasing user motivation with each successful revolution.

The reverse is true as well: Parents and other close caregivers who model an attitude that AT learning and use are difficult burdens, extra work, and more trouble than they are worth in the home setting will likely set in motion a set of user variables that will work to minimize motivation. The initial window of enthusiasm and acceptance of a new device, and the demands it may place on a user to learn, it is often small and narrow. Complicate this initial period of acceptance, and you have possibly predisposed the endeavor to failure; the learning that occurs will be of what the user *cannot* do with the technology, rather than what he or she can and might be able to do. Parents and other caregivers often expect a lot from the interjection of AT into the home and daily life of their user. Although the immediate impact of some AT devices can surely be nearly instantaneous and dramatic—such as the freedom of choice that EC devices can confer in a home environment—more often, the learning and use curves are more gradually sloped. The device takes a while to figure out, and its usefulness in the home environment becomes more apparent only as all concerned become more versed in what it can and cannot do—and more adept in its operation. The immediate and long-range effects of motivation to learn and to use any device well, and to incorporate it into daily life, can be positive results of family attitudes and influences of professionals surrounding a family's introduction to AT use.

Most important among these professionals beyond the home setting for children is usually the teacher. The early-childhood educator, special educator, or other teacher who has primary professional contact with and responsibility for the special-needs child can be a major influence on the child's and family's AT use and success. Teachers commonly are the persons other than parents who will spend the most time with a child. They will rely on the AT devices and systems initiated to help the child participate in a daily variety of activities of communication, learning, self-care, mobility, and other significant life involvements with the child. Teachers and instructional aides are present with the child during many of the most important pursuits of their learner's young life. The motivation to use AT to have fun, to enjoy moving or talking or helping in daily activities, can be significantly reinforced by them. A skilled educator who can model AT success through their positive, can-do attitude and their willingness to learn and grow in applications of a given AT with a child in their daily charge can be among the most powerful influences on what that child feels and knows about what AT can do. Similarly, educators may influence what the child's parents or other close caregivers believe about and do with AT. The influence of the teacher and the instructional aid on establishing and cultivating user motivation is crucial; it ripples through the family, care staff, and potentially all other professional staff who have contact with the child.

Other professionals who can also surely influence motivation of child and adult users, as well as their families and daily care providers, are the specialists from the rehabilitation professions. The occupational therapist, physical therapist, speech-language pathologist, audiologist, rehabilitation engineer, and others who are perhaps in less frequent contact with the child or adult AT user can also be powerful motivating influences and resources. Their incremental task analyses and clinical teaching across an expanding, wider range of component skills can motivate a user to develop initial skills with AT, and to refine more skills to the point that these growing AT use abilities can be incorporated into daily life at home, in the school and community, and at work. Although the contact of OT, PT, SLP, audiologists, and engineering professionals with the AT user and family is often not as frequent or extensive as that of the teacher, the roles of each of these specialized professionals can be key in analyzing and solving initial and ongoing challenges with AT. These professionals can consult with and lend expertise to families and educators as they work with their AT user to successfully implement AT each day in a potential multitude of settings. The special clinical support provided by these professionals can be essential in helping the client or student, his or her parents, family, and teachers acquire, configure, learn to use, and problem-solve to stay motivated for continued use of and success with AT. Though not all these rehabilitation professionals are always specifically trained in or highly experienced with AT, their impact on user motivation to succeed when they do have such training or experience can be large and critical.

In addition to their role in facilitating the motivation portion of Baker's equation, the rehabilitation specialty fields of OT, PT, SLP, audiology, and rehabilitation engineering perhaps play their largest and most important roles in reducing and minimizing the major factors on the lower side of the equation; those are the "load" or effort factors that can individually and collectively overwhelm even the best efforts to foster user motivation.

Each profession can address these effort factors from a different perspective. It is appropriately the role of each to apply their expertise to minimize each of the factors on the lower side of the equation (P, C, L, S, T), allowing the motivation side (M) to become as relatively "top heavy" as possible. The smaller the accumulative load side of the equation becomes in relationship to the motivation side, the more likelihood there is of AT success occurring with a given user.

So while parents, family members, caregivers, teachers, and instructional aids surely have key roles in establishing and fostering motivation on the top side of the equation, the habilitation and rehabilitation specialists have the key roles in helping to minimize each of the areas of load or effort on the underside of the Baker equation as related to their individual areas of expertise. Let us look at some of the specific roles these professions play in reduction of effort.

Occupational therapy (OT) and physical therapy (PT) professionals can analyze a client's needs, and work to reduce a user's physical effort by addressing seating and positioning adjustments, range, and extent of motion for control site and switch use, strengthening and developing accuracy of movement for control sites to use with AT, and intervening with device configuration and client training to help minimize complexity of movement sequences needed to operate AT devices. OTs and PTs contribute valuable expertise in finding control sites on the AT user's body with which the student, client, or patient can reliably operate the switches and controls of their AT devices. They can help the user develop the strength and accuracy of these sites, whether hand, finger, foot, knee, head, elbow, or other body part/movement, that is needed for optimal accuracy and efficiency of use. These professionals can help make decisions and recommendations to all the others involved with the AT cadre on where to locate and mount switches and controls for a user. As the user then repeatedly activates and operates devices over time, his or her activation movements can be done with the least exertion and the least possible wear and tear on limbs, joints, and other body tissues. OTs and PTs play key roles in assuring that the AT user is seated in their wheelchair or other supported environments/devices in ways that are comfortable and that allow for optimal postural security. Postural security ensures that the user's hands and limbs, head, or other control sites are aligned and balanced for optimal use in operating devices. This optimization occurs while reducing user exertion and discomfort across all uses and situations and positions, whether the user may be sitting, lying, standing, or in other positions.

Speech-language pathologists (SLPs) strive to enhance the linguistic abilities and efficiency of their client, patient, or student, and to minimize linguistic load on the AT user. Specifically, this work by SLPs may include therapies and instruction, in conjunction with educators and others, to improve all aspects of the client's literacy abilities: speaking, listening, reading, writing, and thinking (Montgomery, 1995). SLPs also analyze communications tasks and work to help reduce the amount of load placed on the AT user related to representation of meaning by pictures, symbols, words, or other symbolic systems required to use a given assistive device. SLPs are particularly involved in AAC to teach their clients ways that meaning may be represented (pictures, symbols, words, manual signs, Morse code, finger spelling, etc.); ways to transmit information via visual, auditory, and tactile channels; and methods that can be used to select and indicate representations of meaning. These components of AAC use are addressed by the SLP in conjunction with the client and family, as well as his or her educators and other professionals, to achieve more efficient rates, more suitable semantic content, and greater communicative competence. These areas of communicative competence may include linguistic, sociolinguistic, strategic, operational, or others (Light, 1989) for the AAC user's effective spoken, gestural, and written expression. Speech-language pathologists also may assist in enhancing the user's linguistic foundations for adapted computer access for speech, writing, calculation, and other expressive skills. This can be accomplished through directed instruction and experience in these areas both in the clinical setting, as well as through extended assignments to be carried out at home, in the school or community, or in work-related settings.

Audiologists (AUDs) are often uniquely qualified to evaluate for and provide clinical leadership in addressing a potential AT user's need for assistive listening devices, and various types of endosomatic techniques to enhance receptive and expressive communication such as signing, finger spelling, cued speech, and speech reading. These latter techniques, though unaided and not necessarily technology-based, are important parts of the larger area of auditory rehabilitation, in which AUDs are the professionals with primary expertise. These endosomatic methods and techniques can provide highly useful support in making the user a more competent communicator and can be taught and used in conjunction with or apart from various assistive listening devices such as auditory trainers, hearing aids, and electronic vibrotactile devices that may be recommended by the audiologist.

Rehabilitation engineers (REs) may help all of the professionals described above, and help evaluate for, recommend, locate, and configure essentially any assistive electronic, mechanical, electromechanical, hydraulic, or any other helping technology for use by persons who have special needs. RE professionals include engineers and technicians whose knowledge and expertise in how to design, fit, modify, and optimize a variety of technological devices and systems for specific users with special needs can make the difference in whether these devices or systems work for a given client or student—or not. RE professionals are still few, but their impact in the AT field of practice, across all component areas of AT, is vitally important to users, families, and other allied rehabilitation professionals who depend on their expertise.

100 words about *NUDGE THEORY*

The middle line on the highway forces us onto the proper side. Yes? Not really. It nudges, suggests the correct lane but is not a physical barrier. Rather than punishing or penalizing for behaviors, Nudge Theory offers design suggestions of where to stand, drive, what eat or do by indicating safer, more attractive, more available behaviors. So, put that bowl of fresh apples on your front table, with the candy bars on a higher shelf. Nudge Theory encourages people through subtle incentives to make decisions and behave in ways supportive of their self-interest without threatening immediate negative consequences. See Thaler and Sunstein (2008), *Nudge: Improving Decisions…*

Also check https://suebehaviouraldesign.com/nudging

SUMMARY

Issues surrounding human factors are closely linked to success and failure of AT implementation in daily life, school and work settings. Baker's equation depicts and summarizes some of the major, more global considerations in AT use. It can be understood from a user's perspective of motivation and effort. It can also be an organizing framework for viewing the roles of the variety of related education, special education, and rehabilitation professionals who may practice in the field of assistive technology. The assessment, evaluation, and diagnostic abilities of these professionals, plus the nature and intent of the specific types of AT intervention expertise they can provide, all can be more fully understood and interrelated via Baker's equation. Some specific, essential human factors concepts that are integrally interwoven with the more global framework for AT success introduced by Baker's equation will now become our focus.

REVIEW QUESTIONS

1. We have listed tech goals from a variety of sources. Which do you believe may be the most valid and important? Which are the least valid and important? What additional goals would you generate? List them.
2. How might goals for a given consumer, client or student be affected by his or her age and/or ability levels? Which goals might become more and less important across these variables? Why?
3. Should AT goals be written for and include parents, families, care providers, and education/therapy professionals? What would these goals be? Why?
4. Without looking back in the book, write or draw out the Baker Basic Ergonomic Equation for motivation vs. effort factors. Describe each of its components in your own words. Now look back at the book. Did you recall and reproduce it accurately? Fix up your answer if you need to once you have looked at the equation again.
5. From your professional and personal perspective, what would be your own role in helping to maximize an AT user's motivation? Describe five specific ways.
6. Again from your professional and personal perspective, what would be your role in addressing each of the effort or load factors presented in the Baker equation? Describe three specific ways for each load factor (P +C+S+L+ T).
7. From your understanding of general consumer technology as well as AT, do you think the Baker equation is a valid representation of how humans interact with devices? Specifically why? Or why not?
8. In addition to the AT user, who is the next most important person or professional in maximizing AT success? Why?
9. What role might society in general, a user's cultural differences from mainstream society, religious factors, or other such variables play in the motivation-versus-effort equation?

10. Who or what might be primary detractors of motivation for a child or adult? Why?

5

REFINING HUMAN FACTORS IDEAS

The best, most expensive and elaborate consumer and assistive technologies, coupled with the best, most learned professional work of all of those whom we have described so far can be for naught if essential human factors in assistive technology are not considered. Variables that surround the other *people* in the user's life, including issues dealing with professionals and cultural differences, may have an impact on the successful use of AT by a consumer. So might variables that surround the users personally and the AT devices themselves (see Chapter 9). Global factors such as culture, ethnicity, religion, gender, and age also may affect how an AT user, as well as his or her family and community, accept or reject the entry of assistive technology into their lives. Some experienced professionals and users in the field of AT have stated that perhaps as much as 75 percent of the eventual success of new technology use and intervention for any given client eventually focuses on how they perceive, relate to, and interact with the enabling devices in their lives—that is, on human factors in the technologies they use (Anderson, 2021; Helander, 2006; Denning, 1996; Botten, 1996; King, 1999).

In this chapter we will cover ten specific, essential areas of human factors that can be applied to consumer devices, and to our practices in assistive technology. These human factors are critical in assuring that the AT interventions we investigate, select, and acquire for our students, clients, and patients, regardless of the specific tech classification, will be as effective and as suitable for the user and family as possible. In previous chapters, we discussed what human factors in AT are and are not. We will now explore some specific, essential human factors, and how we might include consideration of these in our tech decisions and practices, whether our interests are with general consumers, education, rehabilitation, or other areas.

SPECIFIC HUMAN FACTORS IN AT

The ten specific human factors we will review and discuss in this chapter have been selected for their relevance and importance to AT success. They are adapted and extrapolated largely from Norman (1988 and 1993), who wrote of the psychology of everyday technologies, though not directly about assistive technology. Specifically, these human factors areas include the following:

1. *Transparency-translucency-opacity of devices and tools.* This refers to the practical user-friendliness and visibility (or lack thereof) of actual AT operation and use for any given device. We will present some additional thoughts and approaches to these same central concepts we have introduced previously.
2. *Cosmesis of AT devices, tools, and systems.* This means how AT devices and their controls or switches appear to the people who use them and to those around them who interact with the AT user and their devices. Are these items cool, O.K., lame or dorky as school-age users and their friends might say?
3. *Mappings of AT learning, use, and operation.* The emphases here are on natural and unnatural mapping patterns (movement sequences for actual use) that we experience with general consumer devices and with the design and use of AT. Mapping-related concepts of structure, organization, groupings, and direction as related to control and operation of devices and systems are important factors to consider.

4. *Affordances.* This term refers to how the design of the item and the types of materials used in constructing AT devices influence users and those around them. Is there a "psychology of materials"? Does it influence what we think about ourselves or others who use things made of these materials? Do we pull or push on a handle? Does something made of metal afford itself to different uses than the same item made of wood?
5. *Learned or taught helplessness.* These are terms we often hear in relation to those with disabilities and special needs. Could it be that technology use—and barriers to it that we may be unaware of from the user's perspective—can sometimes (perhaps more often than we realize?) be the inadvertent catalysts that bring about the learning (and our unintended teaching) of helplessness for some potential assistive technology users?
6. *Feedback from switches, controls, screens, and devices.* This feedback lets us know we are working as a system with the device—that we are operating it successfully, or not. Feedback can include audible, tactile, visual, olfactory, kinesthetic, and other sensory means to indicate device response. Prompt, accurate, multimodal feedback can be particularly important in AT considerations for persons who have sensory, motoric, cognitive, or other limitations and special needs.
7. *Knowledge of technology use that is "in the head" versus "in the world."* This means the operational knowledge for devices and technology that we carry with us and that is based on our prior learning, exposures to, and experiences with other types of devices and AT, as opposed to knowledge of how to use or operate a given technology that is built into a device or displayed as instructions on the surface of the device. Do the natural mappings of the device or the symbolic information displayed on it tell us how to use it, or do we know already? In short, where does device and tool transparency reside? Is it in our learning and experience, or in design, configuration of and overt instructions for use

with an assistive technology? Norman (1988) and Romich (1994) have discussed this concept at length.
8. *Constraints of AT use.* What physical, sensory, semantic, cultural, logical, or other factors may appropriately limit ways that technologies can be used? Can we know these before we use a technology—or can we discern them from our initial attempts to use a device?
9. *Incorporation of "forcing" or fail-safe functions for systems.* These may include the constraints mentioned here, plus other techniques and methods used in creative ways to increase user and AT device efficiency, as well as to prevent misuse of a device, or injuries and harm to the AT user and others.
10. *Prevention of errors, mistakes, miss activations ("miss hits") in AT use.* This also includes enhancing the likelihood of correct, accurate activations and device use by attention to many or all the foregoing essential human factors.

These ten major areas of human factors in assistive technology, adapted extensively from Norman (1988 and 1993) and King (1999), are among the more important for us to address in this text. Nonetheless, they surely do not constitute the entirety of the human factors field nor the full extent of human factors considerations in assistive technology. They are starting points from which to build our understanding of human factors in AT. We have chosen to deal with this finite set of human factors because they constitute, in our experience and belief, some of the most important areas for focus for a variety of professionals and others involved with AT. Educators, special educators, and clinicians of all types should be aware of these factors at the preservice training level, as well as at the level of in-service professional development. Let us look at each of these areas of essential human factors in much more detail and depth.

Transparency and Visibility

This basic human factor in AT seems so obvious that one may wonder why we spent much of the second chapter on this topic—and why we are now addressing it again. The rationale is simple: How transparent-translucent- opaque a device is—or appears to be—for use in practical tasks directly affects our likelihood of using it successfully. The transparency-translucency- opacity scale for device use can be an even more significant ir\fluence on persons who may not be as able to see, hear, communicate, move, or solve problems as well as others of us. Other terms that we hear related to this central accessibility concept include *user friendliness, openness,* and, from Norman (1988 & 1993), *visibility* of a technology. To those of us with normal-range sensory, motoric, cognitive, and linguistic abilities, the technologies that we find translucent or opaque are probably higher technology, more complex devices that we must spend time figuring out, and for which we probably must receive special instruction and training in order even to begin to use.

Windows, for example, as well as complex word-processing, graphics, or spreadsheet utilities, are technologies that are closer to the opaque end of the scale for many users. So are technologies such as flying a plane, programming a DVR, thermostat, or clock radio, using a graphing calculator, or conducting checkbook transactions via a touch-tone telephone. These are not necessarily highly complex activities for all users, but exactly how to do them is not obvious; they do require some instruction and practice, and they involve the high likelihood of "repairs" of unsuccessful attempts to use the technology. These are opaque technologies at the outset to most users who possess normal-range abilities across the areas mentioned.

But what about using a cup or a spoon or a pencil or a toy car—especially if you are viewing and attempting to interact with the world from a body that does not have the abilities presumed for others? What if you have not experienced drinking from a cup, or feeding yourself with a spoon, or pushing/ pulling a pencil or crayon on paper to make coherent marks, or making a toy car move with your hands and making car sounds using your own voice? These transparent (to the rest of us) devices become opaque or at least translucent to the special-needs user: Some special demonstration, or maybe even instruction and training in just what to do and how to do it with each of these seemingly simple, transparent devices must be provided for these users. Expecting a person who may never have drawn with a pencil or played with a toy car while making motor sounds, or cooked lunch on a stovetop with lots of controls, to decipher how to use these seemingly transparent technologies is not unlike the skilled pilot or computer "jock" expecting the new student in either pursuit to intuit just what to do with these more complex technologies.

It appears then that transparency-translucency-opacity of whatever tools, devices, or technologies we are addressing is really a matter of perspective: What is transparent for me may be translucent or opaque for you, and vice versa. As we have mentioned earlier, based on Sanders and McCormick's (1993) wise insights about technology, we must not fall into the trap of thinking that our own perspective on human factors is generalizable to others. How we see things, how we set up and interact with devices, especially as normal-range users of technologies, may not be at all how our special-needs clients and students view those same devices. The chasm of opacity that can lie between user and use of any given device may be as huge as it is invisible. The wise professional knows this going into any AT evaluation or intervention situation with any potential AT user and strives to make the intended devices as transparent as possible through design, configuration, selection, and addressing of other variables that approach use and learning of the AT from the user's point of view—not the clinician's or general consumer's perspective. Training, prior, and current experiences, motivation, device design, and configuration can all work together to make what was opaque into something that is now transparent—or at least translucent—to the user. Professionals who work in any of the component areas of assistive technology can help orchestrate this more harmonious blending for many AT users—if they know that overall device transparency, or lack thereof, is something relative to each user, and something that can be influenced by allied professionals in the many areas of AT practice.

Cosmesis

We all are concerned about how we appear to others. From early elementary school years onward, our looks, haircut, clothing, makeup, accessories, the car we drive, and most other things that are positioned or worn close to our bodies tell others (and ourselves) about who we are, how we think and act, and what is important to us. The appearance of these things in terms of attractiveness generally become increasingly important to nearly all of us as we grow from early childhood into our teen and adult years. These appearance- related concerns are known as the human factor of *cosmesis.* Cosmesis (from the same root as *cosmetic)* concerns how we appear to others in terms of acceptability or "coolness," and how we feel about how we appear to others. With assistive technologies, despite the learning effort and expense that often underlie acquisition of AT devices, devices that do not reach at least a modicum of cosmetic acceptance for users and communication partners will likely be rejected by their users, and will end up unused, sitting in the closet or under the bed. It happens all too often. Inappropriate cosmesis for a given person and a given AT device spells a mismatch that can waste thousands of dollars and many hours of time and travel for the user and family, professionals, and equipment suppliers.

From perhaps about the age of 8 or 9 years (approximately third grade on), most boys and girls in Western society become highly attuned via the media to how they themselves and others appear. The image role models for children, teens, and adults are now essentially omnipresent and are conveyed via television, videos, magazines, movies, and other globally available sources, as well as interpersonal sharing among children, teens, and adults who are friends, classmates, co-workers, and neighbors. When one is expected to rely on an assistive device at any age beyond perhaps about 8 or 9 years, how it will appear in conjunction with the user's body may become a major consideration (Beck & Dennis, 1996; King, Rosenbaum, Armstrong, & Milner, 1989).

Several studies in the augmentative and alternative communication (AAC) component area of AT have shown that families and users of VOCAs (voice output communication aids) may be concerned about how the AAC device appears in conjunction with the user. Blockberger, Armstrong, O'Connor, and Freeman (1993) studied children's attitudes about a nonspeaking child who was using three different AAC techniques and devices. They hypothesized that a child using an aided electronic communication system would be viewed more positively by other children of about the same age than the same child who was using a nonelectronic aided system or an unaided AAC method (signing and finger spelling). Results of their study did not support their hypothesis. They found that children's differentiated reactions to the child using the three AAC methods were statistically insignificant. They concluded that, based on their study, attitudes toward a child AAC user will not automatically improve through the provision of an electronic VOCA to that child. They state that many unanswered questions remain in this area, and that clinicians should counsel parents and other important adults in the child's life as to the possible impact of using technology on how others might view their child.

In a somewhat similar study, Gorenflo and Gorenflo (1991) examined the effects of information given to the communication partners and the type of AAC techniques used on the attitudes of partners toward nonspeaking adult individuals. Results of their work suggested a positive impact on the attitudes of adults toward another nonspeaking adult who used an aided electronic communication device.

The reactions of fifth-grade males to videotaped segments portraying a nondisabled, peer-aged male using each of 13 different adapted computer and AAC access methods were studied by Herbenson and Sather (unpublished research, 1996). They found that their "AAC user" was rated differently based on the type and configuration of adapted-access technique. Their subjects rated the user for each technique on the following scale: **1** = really dorky, 3 = kind of dorky, 5 = O.K., 7 = pretty cool, 9 = really cool. Their results, though tentative and based on small sample size, suggested that fifth- grade males can and did differentiate relative cosmesis among the techniques portrayed. In general, results of their study suggested that access methods that focused more on use of switches and control sites located nearer the face or head of the user tended to be rated lower. Adapted-access methods that more closely approximated "normal" one- or two-hand use of a switch, joystick, or keyboard tended to be rated higher.

In all, studies exploring cosmesis with other component areas of AT and with older individuals are scarce in the AT literature. General consumer advertising by companies such as Target Stores, McDonald's Restaurants, J. C. Penney, Wal-Mart, and others have been showing persons in wheelchairs or using sign language as part of their merchandising models for several years. This trend seems to be on the increase, and it is hoped, will have a sizable impact on what we all view as acceptable, even attractive, regarding use of enabling devices, equipment, techniques, and clothing. The types of AT depicted in these ads, as well as in the internationally televised Paralympic Games, most marathons, some ski races, and other adapted athletic events, may also serve well to acquaint the public with some of the more overtly useful assistive technologies and with the dedicated, even courageous users who have become so skilled with them. The various types of color-coordinated racing wheelchairs, aerodynamic suits, other special equipment that appear sleek, modem, and professional may do much to enhance the public's view of what is fashionable and "in," even if it is part of a person's AT gear.

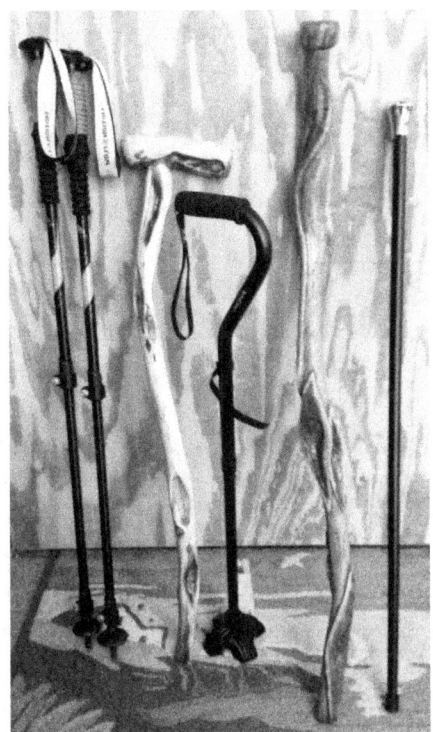

Cosmesis and satisfactory, reliable function are important even with simple devices. Which of these common assistive mobility items appear more appealing for general social use? For rugged trail use? For theatrical use? Which would you choose for daily use if it worked well for you? Or for looks, alone? Why?

Mappings

Whenever we use a device, we typically have specific expectations of how we are supposed to move the device itself, or how we are to move the controls and switches, in predictable, discernible ways that allow us to operate the device. The directions and extent of these movements in a device that is designed well will follow natural patterns called mappings that allow us to decipher easily how to operate the device. We expect certain types of results and occurrences from certain types of input to the controls of tools and devices we use. These mappings can flow in natural, evident, expected ways based on how our bodies naturally move and how the world seems to work naturally. These natural aspects of mappings can be expressed in terms of what we intuitively know, or from what we have learned through experience or instruction with aspects of our physical and sensory universe, such as gravity, locking and closure, speed, intensity of stimuli, coolness or warmth, and many other aspects of our lives and environments. In technological society, we may wish to modify any of these factors and more via use of devices that require mappings—preferably natural ones.

For example, a powered wheelchair controlled with a hand-operated joystick generally is configured to move forward when the joystick is pushed forward, backward when the joystick is pushed backward, and so forth for the other directions in which the chair can move. In a well-designed powered wheelchair, the rate of acceleration and speed of movement of the chair is proportional to how far the joystick is advanced in each direction, and to how quickly the joystick is pushed that way. The more quickly the joystick is pushed, the more quickly the chair accelerates; the farther the joystick is pushed, the faster the chair moves. In an extremely well-designed chair, backward movement of the chair is at only a portion of forward acceleration for obvious safety reasons—the chair must not jump backward as quickly or accelerate to as fast a speed as it can travel in a forward direction.

To try a thought experiment with mappings as described in this case, imagine reversing the connectors from the joystick control so that the chair operates in the opposite manner across all directions. That is, if forward is backward, backward is forward, right is left, and left is right! The confusion this would cause is readily apparent in this thought experiment; unnatural mappings in this case would make this AT device (the powered wheelchair) almost impossible to operate easily and safely, especially for someone with cognitive or motoric limitations.

Other examples of natural mappings include switches and controls that allow minimal flow of electrical or water current when we first turn them "on," then allow greater flow (greater intensity, volume, loudness, etc.) as we continue to move the control through its arc of movement, or as we push or hold it longer. In the United States and Western societies, the mapping for most of these types of operations is to move a control or switch increasingly up or to the right for "more," "louder," "hotter," "tighter," "faster," and so forth. We typically move controls or switches to the left or downward for "less," "softer," "cooler," "looser," and eventually for the resting signal (often a click or beep) and the stable position indicating inactivity, or more simply, "off." We expect most devices to operate this way; whether they be radios, jar lids, thermostats, faucets, bolts, computer switches and on-screen control panels, or any other tools and devices we may employ in our daily lives. If for some reason, such as poor design or malfunction, devices do not work in this manner with these relatively natural mappings, then we find that the translucency, even opacity, of device and tool use increases substantially for us. The introduction of unnatural mappings into the operation of technologies makes them more difficult and confusing to use.

To illustrate further with some examples from adapted computer access, environmental control, and AAC, consider the following actual situations that the author has observed with AT users over the years. One person living with disabilities who used various adapted methods of access to his computer had a phone directory set up in the computer's memory that he could use to retrieve and dial calls. His teacher had helped him set it up at school several years before, and he had added to it over time. The problem was that the directory was set up essentially in random order. Using the more natural mapping strategy of alphabetizing the list helped this adapted-access computer user to store and retrieve information more quickly and effortlessly than with his prior random system. This may seem like a small convenience to many of us, but when you consider that this person's keystroking rate was about 10 or 15 characters (two or three words) per minute, reliance on a more natural mapping strategy to help him store and get to information he needed saved him much time and effort over the many times per day he accessed this now alphabetized list.

In another case, a child's beginning AAC symbol board eventually grew to contain the vocabulary items of *up, down,* and *under.* The way the board had evolved over time, however, with symbol vocabulary items being added on as the user learned them, *up* was placed inadvertently in the middle of the bottom row of the 4X8 row-column symbol matrix, *down* was two items over from up, and *under* got located in row 2, column 8, above and to the right of *up* and *down.* Confusing? Yes, indeed! The board had simply evolved with new items as the user learned them, and his SLP and teacher placed them at distinct locations on the board to aid his initial locating of them. But despite their initial rationale in trying to "spread out" vocabulary items so they could be located more easily, they ended up creating symbol selection possibilities that did not follow natural mappings—especially as the young user became better acquainted with just what each term referred to in the external world. Can you guess how the symbols were repositioned when this was pointed out to his clinician?

Another simple yet important example of inattention to natural mappings and the confusion this can cause involved a young quadriplegic adult man who had suffered a C-2 spinal cord injury. He now capably used a powered wheelchair controlled by head-mounted switches and by using a puff and sip switch, and lived largely on his own with minimal assistance in an accessible apartment. He used his puff and sip capability also to operate his XIO environmental control unit. This device would scan an indicator light down a menu of eight items he might wish to turn on/off or open/close. He would puff an entry code to start the XIO device scanning, then puff again to select the appliance or other device (television, lights, radio, etc.) located around his living room and bedroom that he wished to operate. The devices that he could control via the scanning EC unit were arranged on the scanning menu in random order and were not "mapped" or organized in any particular pattern according to placement in the room or order in which he might want or need to operate devices for convenience or safety. So he commonly was turning on the wrong light, or inadvertently opening a door, and having to wait until the XIO scanned to the item he wished first to operate upon entering his living room or bedroom (typically a main floor lamp in each for maximal illumination and safety).

The mapping we rely on for such devices either places the items left to right in the rooms we enter in a top-to-bottom arrangement on vertical display, such as on the XIO, or places the items in priority order of operation in top-to-bottom configuration, with the highest priority, most desired items at the top, and the lower priority, less desired items toward the bottom. Reconnecting of this young man's XI0 and the wireless appliance modules used with the devices around his home in a manner that involved natural mapping of left-to-right items in each room to correspond to a top-down display on the EC scanning menu helped him to make faster and more accurate selections and increased his motivation to continue using and his satisfaction with the XIO EC device, which gave him greater control over his living environment.

Affordances

We are influenced by the type and composition of materials that surround us; especially the things that touch us and that we touch. From the luxurious feel of "rich Corinthian leather" to "the touch, the feel of cotton," to the solid feeling that cold blue steel imparts, there indeed appears to be a "psychology of materials" that can affect what we do with—and what we think about—the materials, substances, and composition of items around us (Norman, 1988 and 1993).

We are also influenced by the design of things, and how they appear. A large metal plate on a door front says "push" to us, even if no words are present; a round or handle knob says "turn and push"—generally, but not always. A picture printed on a paper page says "touch"; the same picture mounted on a plastic plate switch tends to say "press," and on so on. These few examples illustrate what are known as the *affordances* of device designs and materials. That is, what uses are afforded by the design and/or composition of a device with which someone comes in contact (Norman, 1988 and 1993).

In the case of AT, wherein the user is often closely linked, even literally harnessed, to the devices he or she requires for access to the world, the materials that touch their body have much to say to the world—and to the user— about who the user is. The AT user's opinion of her- or himself is also affected by the affordances of the substances, materials, fibers, and other physical aspects of the things that he or she touches and interacts with closely for much of each day. Since the mid-1970s we seem to have discovered color and consumer appeal in many of the items used by all people, including those living with disabilities and challenges. It used to be that a wheelchair was made only of shiny chrome-plated steel tubing, with slick plastic or leather sling seat and back stretched over the frame. The mere appearance of this type of chair, which you still see around quite often, seemed to shout **HANDICAPPED** or **INSTITUTIONAL** to the viewer and the user. Now, many wheelchairs are available in a variety of colors both for the frame and wheels, as well as for the upholstery. The use of color-coordinated anodized aluminum and other alloy tubing, combined with matching and/or complementary-colored upholstery material for the seat, armrests, and back support, speaks much more of home or office furnishings or clothing accessories. Wider incorporation of softer fabrics and materials, color-matched metal frames and seating inserts, and more stylish overall designs have done much to make wheelchairs more appealing, common parts of everyday fashions and furniture, and less stigmatizing, less segregating for the user.

Similarly, augmentative communication devices have taken on a more appealing, consumer-friendly appearance in the past few years. AAC devices used to look like a cold metal and plastic box, generally gray or beige, with an array of switches on top. Since the early 1990s, the appearance and appeal of AAC devices has changed. Now, many of the devices on the market from the 1980s and well into the 2000s, such as the Prentke Romich Company AlphaTalker and DeltaTalker, as well as the Zygo Industries Blue and Green Macaw series of devices come in a variety of bright colors. These devices, as well as the DynaVox II by Sentient Systems Inc., the Pegasus by Words + Inc., and several other assistive communication devices made by other manufacturers are designed with smooth, rounded, compact lines, and more eye-pleasing colors and designs than the gray or beige boxes of the past that seemed to give a laboratory or institutional aura to the device—and its user. AT devices and equipment, like so many other things in consumer culture, have become attractive, inviting, and acceptable in design, color, and overall appearance. They afford inclusion, acceptance, and even fashion appeal to the user, rather than isolation and segregation.

Many strides can still be made to transform the mobility and communication aids, as well as other AT devices and equipment, that many of our students and clients rely on daily into items that are as acceptable as a personal headphone stereo player, phone, tablet, or a set of designer eyeglasses. The materials and substances from which these evolving designs are made from can afford acceptance or exclusion to the user. What these materials afford for their users tell us and society, as well as those who rely on AT, much about who is competent or not, who is included in contemporary society or not, and who is capable of working and learning or not. Affordances of design and materials used in AT device construction can be major factors in bridging AT users to the inclusion in the rest of life's activities—or major factors in separating them from the world around them.

---

100 words about a practical violation of *AFFORDANCES*

In one of this author's previous office buildings, near a busy bar and restaurant street, a secluded first-floor custodian's storeroom with shelves and supplies had a toilet and utility sink. In a corner, across from the utility sink, was a low, wide fiberglass mop sink for floor cleaning. When the building was open for late meetings and classes, celebrating imbibers often thronged by, especially on weekends, holidays and at semester ends. How do you think the mop sink was used by some reveling visitors in the crowds who knew of the storeroom? You guessed it. A floor-level mop sink affords several creative uses beyond just mopping.
https://www.interaction-design.org/.../affordances

Learned or Taught Helplessness

If nothing succeeds like success, then nothing causes us to feel helpless like repeated failures in our attempts to use devices and technologies successfully, especially items we see others using regularly and well. When we attempt to use a device and cannot, the tendency for all of us is to blame ourselves. We tend to believe, and even become convinced, that we ourselves, not the device, are the reason the human-machine system failed. How often have we attempted to load some new software program, or perhaps to drive a stick shift car in hilly terrain, or to program a new DVR or smart watch and failed? When this happens, we tend to internalize the blame for the system failure, blaming the human, not the machine. The first reaction of most people is to think "I'm too dumb," "I'm too uncoordinated," or "I'm just too slow, inept, or stupid to do this," and, eventually, "Why try? I'll never be able to get this right."

Most of us can tolerate a first attempt and failure with technologies, if the results are not too catastrophic—but it seems that two or more failures to utilize a certain tool or device successfully begin to make us question our competency, motivation, or willingness to try again to succeed. As our failures become more numerous, frequent, negative, or harmful in results (dumping an important program on the computer, deleting a show on the VCR, a car collision, and so forth), our motivation to continue use of a given device declines. As our motivation to try a device again and possibly learn to operate it better declines, our belief that we are the primary contributor to the system failure increases. Soon we have a cycle of learned helplessness in place that might be described in these six steps:

1. I saw or heard others succeeding with this device and using it well.
2. I tried at least once, maybe many times, and could not do /use it properly.
3. I am apparently just too stupid, inept, disabled, uncoordinated, dense (insert your favorite adjective) to use this correctly.

4. I will not try any more for the above reasons. This thing is too hard.
5. This device has taught me that I am helpless to do any better.
6. I am therefore giving up trying. Matter closed. You do it for me.

Most of us in a technological society have probably experienced this insidious cycle of learned and taught helplessness with some tool, device, or type of consumer technology. Now imagine meeting the world each day with limitations, disabilities, or special needs that cause even more of our interactions with technological devices to be unsuccessful—as many of our consumers, students and clients do who may be candidates for AT. The cycle of learned and taught helplessness could likely begin even sooner and take root in us more powerfully because of the multitude of failures we could experience. Although we professionals in the assistive technology fields continue to believe in and promote the central concept that appropriate applications of AT can reduce, even prevent, the development of learned helplessness, especially if we intervene early in the lives of those with special needs, the unfortunate truth seems to be that use of many poorly designed or poorly selected technologies can present numerous significant learning barriers that may indeed cause our customers, students, clients, or patients to experience repeated failures—to be taught helplessness, failure, and incompetence by many of the very devices they are attempting to learn to use to make their lives more independent. The expectations and attitudes that we hold as interacting partners with the AT user can affect this cycle of learning to be helpless. In selecting, teaching, and designing AT, user friendliness, plus early and frequent user success experiences with people and with simple to more complex technologies, should be our intent throughout. Setting in motion a cycle of learned competence and confidence through early successes with easy-access technology may ultimately inoculate an AT consumer for the temporary discouragement later failure experiences may bring.

Here again is the central concept of transparency with which we began this book: The more transparent that technologies can be made, especially for users with special needs, the more likely these users are to employ them successfully right from the start. The positive cycle of use-success yielding more use-more success can be initiated right from the beginning if open, visible, transparent AT interventions are chosen and configured correctly. Parents, educators, clinicians, and others who introduce AT into a person's life can use what they know about AT and human factors to work incrementally with an adult or child consumer toward early and lasting successful experiences, competence, and confidence.

Feedback

A central process in using any tool or device successfully is being always able to monitor exactly what we are doing with it. Imagine trying to type this page without seeing what characters appeared on screen after each keystroke, or without hearing the click of the keys on the keyboard. Or imagine trying to steer a car down the highway or a wheelchair down the hall without seeing the direction changes you caused each time you moved the steering wheel or joystick control. These are just a few examples of the many types of feedback from tools, devices, and technologies on which we rely from moment to moment to use them accurately, efficiently, and safely. Feedback from controls, switches, and devices may take many forms and may combine several modalities. The types of feedback chosen by given individuals may depend on personal preference and experience, on their special needs and limitations, and on what feedback the technologies they are using do and do not provide.

Reduction or loss of one sensory modality through which to monitor feedback from AT may increase a user's reliance on other channels through which to receive this vital information. Feedback through the auditory channel may consist of beeps, clicks, or pitch and loudness changes from devices as they are activated. The array of sounds emitted by most standard computers as different functions are enabled or limitations (e.g., page bottom) are reached are examples, as are the chirps, or auditory read-back of speech emitted by many AAC devices as each keystroke is entered. These audible indicators let users know that they have connected with the intelligent device, and that they have entered the intended item they selected.

Feedback also can be visual. Most users of AT and other consumer technologies probably rely to a great extent on watching which key they are pushing, which switch they are activating, which picture they are touching, and so forth. Being able to see what we are controlling and how we are controlling it is perhaps the most used form of feedback on which we all rely. Although visual feedback can be supplemented or even supplanted to a large extent by other modes, the degree of precision often achieved by it in device use is difficult to replace by other means in tasks where safety, speed, and first-time accuracy of activation are crucial. Some of these tasks may include communication, mobility and transportation, effective operation of power tools, and safe use of cooking apparatus, as well as other technologies wherein mishaps causing physical harm can be the result of inaccurate activations.

Tactile feedback, or how something "feels" as we use it, is an important supplementary type of feedback for sighted persons and a valuable primary source of feedback for persons with low vision or those who are blind. The "crunch" we feel as we enter numbers on a calculator, the resistance and "pop" we feel when turning the on-off switch on many radios, and the locking into up or down position we feel on many common light switches are all examples of feedback redundancy offered to us by tactual cues from common technologies in our environments. Tactile feedback helps us to monitor and control devices when we cannot hear or see them well because of sensory limitations, disability, or environmental interferences. It also helps users with normal-range hearing or vision to tell more accurately what they are doing with a device when they are using it in high-noise, high-distraction, or low-light environments.

Tactile feedback appears to be closely related to kinesthetic feedback; that is, how far up or down, left or right, forward or backward we move a control tells us much about what we can expect the device to do. In general, the greater the range and extent of movement, the greater the range and extent of performance the device or tool will deliver; Turn the radio volume knob farther to the right, and the intensity of sound will get louder; push the accelerator pedal on the car farther forward and the car will move faster; slide the dimmer switch lower on the wall, and the lights will be less bright, and so on. Knowing what the switch or control feels like as it moves through its range and being able to monitor corresponding device response as we move controls through their range of operation are important data delivered to us by tactual and kinesthetic modalities of feedback. We rely on them to augment our visual and auditory methods of gaining information about the devices we are using. We trust them as indicators of what we should be expecting from the devices and technologies we intend to use in practical applications.

---

100 words about professional basketball and *TACTILE FEEDBACK*

The "feel" of an object gives us considerable feedback on how we operate the device. Scoring was down in the 2021 National Basketball Association season. Some believed it was the feel of the new ball. In 2006, a synthetic microfiber ball was tried and rejected after a few months. This season, the NBA again switched to a ball from a different manufacturer. Players soon expressed dissatisfaction, some reporting the ball "stuck" to their hands, impeding their polished timing for shots and passes, and affecting offense and scoring. Many agreed they would just adapt and keep playing. See https://nba.nbcsports.com/2021/11/03/nba-players-struggling-to-adapt-to-new-wilson-basketball

---

Knowledge of Technology Use Within the User or Within the Device

Here's a quick check test. Put your hands on the table or desk where you are reading this book, and "type" out your name, address, and phone number. If you are like most people, even most experienced typists, you will find that the recall of the personal information requested is easy; it is information that you carry with you "in your head," as Norman (1988) discusses so well. On the other hand, the exact movements of your fingertips to "press" the keys on the imaginary keyboard in front of you will be anywhere from somewhat harder to remember to totally impossible for most of us; we simply do not remember where each of those keys is. We rely on the key labels on each keyboard. When typing, we depend on the keyboard in front of us to remind where each key is; most of us do not memorize these locations. Our ability to select the intended keys quickly and accurately is, for most of us, dependent on our ability to see them. Beyond the usual number and letter keys, this is especially true for the function keys that may be labeled and/or placed differently on various keyboards. We may have gotten used to a Macintosh keyboard over the years. Now, when we type on a Windows 95 keyboard, we find that we can touch the letters and numbers fairly well but have to slow down for the function keys and occasionally correct "miss hits" that we make when, by habit, our fingers touch areas of the keyboard to which we have become accustomed for one function, and which now activate another.

The question arises, then: "Where does system transparency reside?" Is it in us? Do we carry with us the experience, prior knowledge and training, or innate intuition to know how to operate most electronic switches and controls, plus most mechanical tools and devices? That is probably not so for most of us. At the other extreme, do we always need to read the full, exact instructions on how to use each hammer we pick up, or how to dial each telephone we use, or how to figure with each calculator we may encounter at work or at the bank? Again, for most of us, probably not. With many devices, their use is so transparent that we can tell from the outset what to do with them. Our experience repertoire and possibly brief instructions displayed in symbolic or graphic form on the device will give us as much background as we need to know with the device. This is information that we do not need to carry with us because it is always available from the device we use.

Not convinced? As general consumers, we deal with many important, common, but different numeric keypads: those on our calculator, remote controller, telephone, and computer. Each is slightly different in number key location; yet we make the transition among them and operate each with little problem because the information we need to use them is clearly labeled right on top of each key. Where is the "7" on your calculator? Where is the "3" on your telephone? Most of us probably cannot remember this information on our own, yet when we have the keypad in front of us our fingers fly to the correct keys with no problem. In these cases, the knowledge of how to use these devices resides partly in us but also partly (and in a transparent way) in the devices themselves.

These concepts become vitally important in the pursuit of successful assistive technology use. How can we presume that even some of the more transparent, visible, obvious tools and devices that we rely on as general consumers will be similarly transparent for someone, young or old, who has not seen or touched a computer keyboard, or a steering wheel, or an array of control switches for lights, stove, or elevator? The important role of early intervention, and of exposing those who may be chronologically as well as developmentally young to correspondingly transparent technologies in the form of adapted toys and other devices, becomes apparent. It would seem to follow that an experience repertoire developed through skilled incremental teaching with and exposure to selected simpler technologies can lead to greater motivation, skill, and success with more complex technologies later on in our students' and clients' lives. Such incremental exposure to and experience with developmentally appropriate toys, devices, and tools can build the user's internal knowledge of how to use technology. Most of us develop this innate ability by play and manipulation with blocks, balls, and other mechanical and electronic toys and items from an early age. For those who have not been able to participate in "moving their world" in these ways, and in learning some basic operational skills and day-to-day physics that are transferable to many other devices, our clinical teaching and directed play with technology must help fill in the missing knowledge from play experiences that never happened.

For more complex technologies, how do we represent and transmit meaning about use of devices to someone who may not share in the common experience, or knowledge base assumed by even some of the more basic devices mentioned here? For persons who have not turned knobs or pushed buttons on toys, who cannot see the keys or hear the feedback beeps from a telephone, or who cannot read the device labels on their environmental control device, how do we convey the knowledge they will need to succeed with these devices? The answer lies in creating motivating learning situations, wherein successful use of the devices in question occurs and reinforces further successful use, and it lies in creating as much sensory redundancy as possible in our incremental teaching and exposure of potential users to AT.

---

100 words about *SHAPE CODING*

Shape Coding offers knowledge from the device, affording immediate tactile feedback. Shape Coding uses specific 3-D forms of switches, tools, and controls so users can identify, distinguish, and verify items by touch without having to look. Alphonse Chapanis pioneered Shape Coding to improve aviation safety, implementing singular shape-coded control knobs, switches and handles in aircraft cockpits. Individualized shapes facilitate pilots' or other device operators' control recognition, further reinforced by proxemic, locational and visual verification. Color, texture, sound, scent, plus sensory and proprioceptive variables can also be useful for coding critical information. See https://www.fhwa.dot.gov/publications/research/safety/98057/ch04.cfm   https://ergonomicsblog.uk/coding-methods

---

Constraints of Tech Use

Physical, semantic, cultural, and logical constraints on how we perceive and use commonly available devices for the general consumer market have been discussed in a detailed, interesting fashion by Norman (1988 and 1993). As extended from Norman's work, the concept of constraints, regardless of the specific area mentioned here, focuses on what we can do and cannot do with various technologies. From an AT perspective, some constraints are obvious; We would not use a dual-handled cup to write a letter, nor would we try to brush our teeth with the footrest from a wheelchair. These examples seem silly and obvious; yet they are examples of primary physical constraints; The devices mentioned simply cannot be used to accomplish the purposes listed.

Most of us "know" this without being taught it on the basis of our general experience repertoire with tools and devices.

But what if you have had no experience or prior learning with cups or footrests or pencils or toothbrushes, and with what these items can and cannot do for us? Then, for you, what constraints (or limits or focuses) would pertain, and what would you believe you could use these tools or devices to do? Once you have tried once or twice to do something with an item and have failed, you may realize the physical constraints inherent in the design of the tool or device—if you have the sensory and cognitive abilities to comprehend the constraints. Some AT users do not have these abilities and may have to rely on their caregivers, parents, teachers, and therapists to perceive and interpret the physical constraints of items they use. For some users, the limitations of device use and non-use may have to be taught incrementally; for others, the physical constraints of assistive technologies in their lives will be obvious from the start.

Semantic constraints deal with the meaning inherent in a device—that is, what we know about the use and intent of the device limits what we do and do not do with it. For example, we understand that a telephone is a device used for voice communication. We could use the handset to drive a nail, stir lemonade, spread wallpaper paste, or crack a walnut, but we don't because we are semantically constrained from those actions by the meaning of the telephone to us as a device of communication.

Similarly, in assistive technology applications, devices designed for one component area are often understood to have function in that area alone—an adapted toy for play only, an AAC device for communication, a wheelchair for meeting basic mobility needs. But what if that toy fireman climbing the ladder could also be started and stopped by adapted switch activation so that he could indicate letters and numbers, or pictures and words as he climbs up and down under control of the child playing (such as with the AbleNet climbing fireman toy, 1996)? A toy turned into an enjoyable communication device—we have broken a semantic constraint! Or what if that high-tech, microprocessor based AAC device that speaks in several possible voices when you press the symbols on the front membrane keyboard array could also be used to turn the lights on and off in your room, or to operate your VCR, or to open and close your doors in your apartment? A communication device used for environmental control— another semantic constraint broken! And what if that wheelchair, which is generally viewed in terms of limitation (we often hear that someone is "wheelchair-bound"), became a lightweight, brightly colored, aerodynamically and ergonomically designed sports machine—a device you could scoot around in quickly and use to race others, beating most two-footed runners? Now the wheelchair has become a competitive, lively mobility enhancement, not a always drag on the user's life. We have redefined the meaning of the wheelchair into a fast, attractive, positive, even "cool" mobility-enhancing device that has lost some of the constraints we had previously placed on it. The semantic constraints of what the wheelchair means to the user and others around him or her have been broadened.

Cultural and logical constraints, the other areas described by Norman regarding technology in general consumer culture, would seem to have much relationship to physical and semantic constraints in assistive technology. In our culture and others, our view of how viable, different, or excludable some people may be linked, at least in part, to what we believe about the assistive devices they may use, such as crutches, wheelchair, speech- output communication aid, and assistive listening devices. It has seemed "logical" to us that someone in a wheelchair could not be an athlete because, until recently in our culture, that is what we have experienced and been told. Further, it may seem "logical" to us that someone using assistive seeing or listening devices may not be a mechanical designer, a watercolor artist, a teacher, or a composer because we have not yet, on a large scale, experienced the work of the many persons in our culture who may indeed be using AT in exactly those ways.

What is logical and meaningful to us about various devices surely is a complex mixture of what our culture teaches us about them and their users, as well as the physical constraints inherent in the design of AT. How we interpret and accept the design and purpose of AT tools and devices across all component areas can be affected by obvious as well as not-so-obvious constraints residing in the devices themselves, and in our vision for and interpretation of intended uses of the device.

Forcing or Fail-Safe Functions

A "forcing function" of a device, according to Norman (1988 and 1993), is a built-in physical constraint that causes the user to operate the device (hopefully!) in the way it was intended to be operated. Forcing or "fail-safe" functions, if they are transparently and well designed, cause us automatically to use technology in ways that are safer, more efficient, and that prevent difficulties, without even thinking about them. Nudge theory also emphasizes users making decisions that will be in their interests and not harmful, per Thaler and Sunstein (2008).

A familiar fail-safe forcing function commonly found in AT used to be the notch on standard disks that forced them to be loaded only one way into the disk drive on nearly all computers. As you look at a disk from the front, top side, each has a small corner cut on the lower left-hand portion of the disk. This cut signals to the user that the disk must be inserted with the comer cut either to the right, with horizontally mounted disk drives, or with the comer cut up into drives that are mounted vertically.

By the way... Why the terms "right" and "up" are chosen as the correct directions in many devices, and why "right" often equates with "up," are cultural and semantic constraints in our society that have led to these directional markers becoming natural mappings for many actions with many devices. Why this pattern has developed and become embedded in our culture probably has to do with our population being largely right-handed, and our linguistic heritage of equating right-handedness and upwardness with righteousness and goodness. But why does it have to be that way? This area of cultural human factors can get complex.

Other forcing functions are found with automatic shut-down circuits devices that reduce battery drain, the speed- and acceleration-limiting features on most powered wheelchairs for backward movement safety, and the numerous redundant command menus and dialog boxes used in computer programs that force the user to make operational selections through the use of discrete-choice, interactive dialogue boxes that will protect their data stored ("Save as," "Do you wish to end this Windows session?," and so forth). Fail-safe and forcing functions, as mentioned earlier, when well designed into devices, are transparent, even invisible to users. They can ensure accurate, safe, proper use of devices of a variety of types. Forcing functions on AT devices must be designed to be of value to the special-needs user as well as the parent, teacher, therapist, or other partner who works with them to learn and implement their devices.

Preventing Mistakes and Errors

We all make mistakes with the devices we use. For those of us with normal- range motoric, sensory, cognitive, and linguistic abilities, correcting those mistakes is something we typically can learn readily; we can independently monitor and accomplish corrections and adjustments with many devices we use daily. This is not necessarily true for an AT user with special needs, limitations, and disabilities.

First, simply sensing that an error activation (a *miss-hit*) has occurred can be difficult for some users. Here is where the role of multiple, redundant feedback capabilities from whatever type of AT device is being used becomes essential. For example, the person with low vision may not be able to read each character as she types it to the screen, but with a speech-readback utility as part of her word-processing program, she will be able to hear and.

it is hoped, detect error entries. Likewise, with an AAC device that has an array of symbols over its activation sites; The speech output from a given activation site can help alert the user and others to whether an accurate selection was made. Tactile markers ("landmarks") can be placed on the selection array to help provide additional sensory data for locating accurate, intended selections and to help reduce error activations.

In the area of powered mobility, examples of preventing errors include the capability of many chairs to be modified in terms of acceleration rate and maximum speed attainable. How fast the chair "jumps" forward or in any direction (acceleration rate) can generally be adjusted to range from very gradual for new users, or persons with considerable spasmodic movements of their control sites, all the way to sudden and rapid for experienced users and those with constant, reliable motor control of the parts of their body they use in controlling the chair.

The concepts of repair strategies and techniques, according to Buzolich (1988), Light (1989), and others, which are critical to successful use of AAC, become important in and can be appropriately extended to all assistive technology. Teaching our students and clients to be alert for and to recognize when communication has not gotten through, when an error on the computer screen has been made, or when they are moving in the wrong direction or too fast in their powered mobility device is the first step. Then teaching them how to correct, fix, repair the error must be part of our teaching and therapy with them. With some users of AT, independent device use and repair of mistakes, errors, and breakdowns (both literal and figurative) may never be possible; their partners, teachers, family, and others must help them monitor and correct errors in AT use. For many AT users across all component areas of AT, learning how to monitor, detect, and fix mistakes and error entries, even despite the many protective forcing functions built into many AT devices, can be valuable areas of instruction and supervised experience for the users and their families, teachers, and clinicians.

This coffee merchant utilizes ordered sets (chunking of bean types) along with positional and color coding, plus mappings, to keep products organized for safety and convenience, and to assure accurate packaging and shipping.

SUMMARY

Each of the ten major areas of human factors in assistive technology that are described here can be interpreted in terms of the Baker equation relating motivation and effort that we described earlier in Chapters 3 and 4. As these factors are addressed by knowledgeable families, teachers, clinicians, and users, they can combine to increase motivation (M) of the user to pursue and complete tasks related to AT; they can indeed all be important, even critical components in successful learning and use of devices. For example, a well-chosen, appropriate AT device of any type that is easy to use and that has appealing appearance to the user and others is far more likely to involve the user in continuing practice and skill development, and to become a useful part of his or her life. The user will depend on the device and recognize its worth, as will the other interaction partners of the user.

As these essential human factors may be ignored or left unattended to, the potential exists for an AT user having to use devices that are too hard to access, are not pleasing or at least acceptable in appearance to the user and others and create more summative load and effort than their limited successful use is worth to the user and his or her family. We tend to abandon and ignore tools and devices that are more trouble than they are worth. Inattention to human factors in our AT selections and acquisitions for our clients, students, and patients can often predispose them to reject what we have labored so hard to prescribe, acquire, and fund for them. We will focus next on some more of the specifics of human factors relative to device inputs and controls that can contribute to AT success when we incorporate them in our planning and interventions—and to AT failure when we omit consideration of them.

REVIEW QUESTIONS

1. Each of the ten major human factors areas we discussed can be identified in many of the common technologies you use every day. Cite an example of each human factor that you have recently encountered (or wished were included) in the daily devices you use.
2. These same human factors are present (or sometimes

conspicuously absent) in the specialized, assistive devices that you may use with your students and clients. Cite particularly good or poor examples of how the ten major areas of human factors we have discussed have (or have not) been incorporated into the AT with which you are most familiar.

3. Which of the human factors discussed so far in this chapter do you think are the most important and the least important in AT? Why?

4. Across the professions of education, special education, OT, PT, SLP, AUD, and engineering, which of the human factors we have discussed may be more in the realm of practice for each of these individual fields? Why? Do these change? How?

5. Kantowitz and Sorkin (1983) have emphatically stated that the first and main rule of human factors is "Honor thy user!" Specifically, what does this mean to you across each of the individual AT component areas and across the different types of special needs presented by your students and clients?

6. What might you do in your field to address some of the specifics you cited? Overall, do you find that you will try to adapt the human to the device, or the device to the human?

7. What should AT consumers of different ages and abilities know about human factors? How should they learn about these? Why—or why not?

8. What error-monitoring strategies do you use? Which are part of the technology you use? Which should be? Why?

9. What strategies could you teach to students or clients to monitor and correct errors with AT devices or other technologies? How would you teach them?

6

SWITCHES AND CONTROLS: YOUR PASS KEYS TO TECHNOLOGY

Give me a place to stand and rest my lever, and I can move the earth.
Archimedes, Greek philosopher Circa 250 B.C.

In his enduring, insightful words, Archimedes was honoring the tremendous power that the correct device in the proper alignment can confer upon the user. In the same manner, switches, and controls, correctly selected, adjusted, and positioned can be powerful tools with which all of us can move our world through use of enabling and assistive technologies. The correct switch or control in the right placement can indeed be the pass key that allows a person to enter the world of independent movement, communication, play, and control of his or her immediate environment—approximating more closely the choices readily available to general consumers of technology.

The wrong control device in the wrong placement (or, worse yet, no control or switch at all) correspondingly locks persons with disabilities and limitations out of the world of opportunities that could await them through independent access to their assistive technology devices and systems. The door to enablement through technology can be securely locked without the proper pass keys to access what lies beyond it.

Throughout our engagement with consumer and assistive technologies, we tend to focus on the bigger items—the major macro-functional devices themselves such as phones, tablets, wheelchairs, AAC devices, EC units, adapted toys and computers, etc. These are larger, more costly, more visible, and typically more dramatic in function, and they tend to be the primary subjects of our efforts in diagnosis, selection, purchase, acquisition, and intervention, versus the smaller, perhaps not as exciting switches or controls that must accompany them.

Despite our wisest efforts in the selection of these larger tech devices, however, the switches and controls that allow our clients, students, clients, and patients to operate their devices for intended purposes are the liberating, enabling pass keys that allow and enhance (or restrict and restrain) users' full access to consumer and AT systems. These "macro" systems may be appropriate, even critical for a variety of important enhanced life opportunities and activities, but if the keys to their access do not fit with the user, the systems can be rendered less effective or even useless.

In this chapter, we will examine and explain some of the essential human factors related to switches, controls, and related interface devices that technology users employ to communicate with and control assistive technology devices of all types. We will attempt to clarify and illustrate some of the essential considerations in the selection, configuration, and use of these switch and control pass keys for full access to consumer and assistive technologies.

SWITCHES AND CONTROLS: DEFINITIONS AND EXAMPLES

What are switches and controls? Strictly speaking, a switch is an electromechanical and/or electronic device that is used to open, close, or modify an electrical circuit. It can permit, interrupt, or direct the flow of electrical current to operate and communicate with a single device or a system of devices (Church & Glennen, 1992; Halender, 2006; Mims, 1983; Schoenfeld, Ed., *et al*, 2022). Some examples of switches in our daily lives include the wall-mounted light switches we all use, the ignition switches and horns on our cars, the activation sites on our smart phones and tablets, and the on-off buttons on our alarms or entertainment devices. Switches surround us in nearly all environments.

A control, strictly speaking, is primarily a mechanical device used to

SWITCHES AND CONTROLS: YOUR PASS KEYS TO TECHN... | 139

initiate, stop, and regulate operation of other mechanical and/or hydraulic equipment. Some examples of controls in our daily lives include our water faucets, the pull cord we use to open our vertical blinds, and the handles and locks on our doors. Controls also are evident throughout our lives in an industrialized society in or homes, cars, worksites, and buildings.

Often, switches may be small, and may involve smaller tapping, pressing, sliding, or flipping movements made from smaller control sites and muscle groups. Switches may also combine electromechanical response via vibration, sound, speech, voice, light, moisture or chemical activation. Controls are often physically larger than switches and tend to involve pushing, pulling, turning, or cranking movements using larger control sites and more powerful muscle groups. In real-world use, the functions of many switches and controls are commonly combined into one knob, handle, dial, or pull cord that we simply call a "control." If you drive a car or fly a plane, you know about combined, multi-factor controls and electromechanical control and switch partnerships.

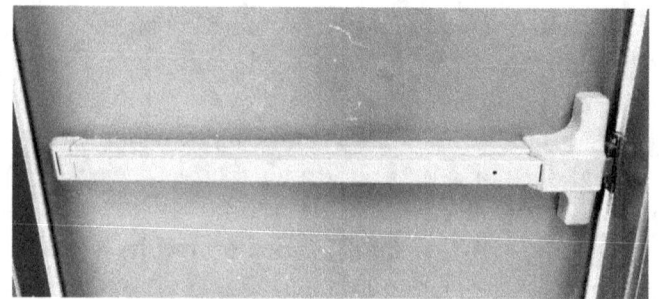

Switches and controls are found throughout our homes, schools, and workplaces. We use them to work, recreate, and adjust our daily environments in a variety of important ways.

Sanders and McCormick (1993) categorized controls of all types into three major groupings: (1) those for transmitting discrete information (on-off toggle switch, rotary selector, etc.), (2) those for communicating and sensing traditional continuous information (accelerator pedal, window crank, steering wheel, etc.), and (3) those for conveying and adjusting information about computer cursor positioning via trackball, mouse, touch screen, track pad, etc.

Distinguishing precisely between what is a switch and what is a control in AT practice may be a pursuit that is more academic than useful in much of real life. Suffice it to say that each could be considered an interface: a mediating device between the human user and the equipment to be operated. Both types of interface devices, switches and controls, as well as ingenious, integrated combinations of the two, are crucial to our access to technology of all types. From opening the front door of our house, to operating a toilet or shower, to placing a phone call, to cooking a meal, switches and controls are the essential pass keys that allow us to unlock the power of other more complex, often physically larger technologies. Switches, controls, and all such types of interfaces are indeed our modern-day levers with which we—and especially consumers, students, clients, and patients with special needs—can move our world.

In industrialized, technological societies, the variety of switches and controls that we encounter as we move through each day and week is vast. In most settings we occupy, we open and close doors, lock and unlock them, turn lights on and off, and operate the numerous switches and controls on our vehicles to drive them and make them comfortable. We all use dozens of other switches, controls, and interfaces throughout each day that we may not even be aware of across many settings. The essential function of these is to allow us, the users, to communicate with the device or system we wish to interact with, operate, or control. All types of switches and control devices allow us to send information to a system and to receive information back about on-off, loud-soft, dim-bright, lower case-upper case, and many other aspects of device functions. It could be accurately said that every switch and control is also a visual, auditory, and/or tactile display device. These devices generally indicate or display to the user what the status of their interaction with the control or switch is by its position, illumination, clicks, and other types of feedback. These indications from the control or switch display and convey back to us, the users, that we are successfully or unsuccessfully communicating the information we intend with the device (Kantowitz & Sorkin, 1983).

Successful access to and use of switches and controls in all our environments are essential to successful living in our society. Reduced ability to operate switches and controls successfully, due to limitations of the control devices or of the humans wishing to use them, across the many settings in which they are essential to our daily lives, can constitute a significant barrier to full participation in much of what life has to offer for residents of industrialized societies.

As an informal experiment and an important learning experience, try this simple exercise: Enter the front door of your home or apartment, and count the doorknobs, locks, and other controls you must manipulate to get in. Then tour systematically throughout your dwelling from room to room, hallway to hallway, while you count every switch, control, and combined switch/control you can locate. Count each item on all devices, appliances, and machines in your home. Do it now. We will wait here.

Done? All right, what did you find? If your home is like most in the United States, and if you counted every switch and control on every wall, smart speaker, stove, radio, DVR, shower, toilet, CD player, tablet, car, mixer, and so forth, then you counted at least 200 to 300 separate switches and controls with which you potentially interact every day, and possibly many more depending on your home and your lifestyle. As some have found who have tried this experiment, you may have gotten to about the 200 or 300 item count and then stopped because there were so many more switches and controls left to count in your home. A daunting task, to be sure, to get every possible one of them. Have you thought of other switches and control devices in your home that you may have overlooked now that we have introduced this topic?

The point of all of this is simple: We rely on the switches and controls around us daily and from moment to moment to interact and communicate with our environment in many settings. We tend to take switches and controls for granted. They simply become buttons, knobs, sliders, wheels, and so on—and they often become functionally invisible to us. We tend to think of the end products of the technologies we use (talking with someone at a distance over the telephone, preserving our thoughts on paper, etc.), and overlook the pass keys and the devices that allow us to accomplish these things. It becomes even more apparent to us through this brief experiment how inability to use these many switches and controls successfully at home and in other settings can be a very real disadvantage in our society. Difficulties with switch and control design, selection and use can constitute a significant barrier to meaningful interaction with the world around us.

100 words about *ZOMBIE FINGERS*

Some users of capacitive touch screens on phones, tablets, and other devices experience "Zombie Fingers." They press on desired activation sites of their visual display but cannot make the device respond. Although often thought to be related to skin temperature of the users' fingers, especially for older or exceptionally thin people, the phenomenon is due more to moisture sensed by the electronic field emerging from the touch screen. Gloved or calloused fingers can be zombifying and yield the same results. Keeping skin moisturized or using a stylus can help. Some users report employing a snack sausage. Really.

See https://www.consumerreports.org/cro/news/2015/06/zombie-finger-and-touchscreens/index.htm

https://royceonarampage.wordpress.com/2019/02/23/the-zombie-finger-survival-guide

---

SPECIFIC HUMAN FACTORS REGARDING SWITCHES AND CONTROLS

Several specific, often invisible factors need to be addressed in the switches and controls that we may consider, select, and modify for our students, clients, and patients. These factors include the following:

- Sensitivity
- Selectivity
- Resolution
- Contrast and shape coding
- Feedback
- Latching and variability
- Positioning and mounting
- Composition and construction
- Cosmesis
- Tactile or sensory defensiveness of user

Attention to these basic human factors surrounding switches and controls can help make the pass keys to our technological society more available to everyone, especially children and adults with limitations and disabilities who are our students, clients, patients, and family members.

Sensitivity

Sensitivity is one of the more common and probably more obvious human factors. We have probably all experienced its presence or absence in the many switches and controls we use each day. Specifically, sensitivity means how much or how little muscle force is required to activate any given switch or control. For example, we all have encountered light switches that were difficult to flip on, computer keyboards that had a stiff feel when we typed on them, or a radio tuning dial that was so hard to move that it seemed it could never be put on to the correct frequency unless we used both hands and our full attention to tune the radio. Switches and controls that require too much muscle force (i.e., that are not sufficiently sensitive for the control sites and muscle force levels used by most people) are encountered all too frequently: the door handle we just cannot turn with wet or arthritic hands, the crank on the boat winch that is nearly impossible to rotate, the hose faucet handle on the outside of the house that neither you nor your burly neighbor can turn by hand.

These and other examples of switch and control sensitivity mismatches for many users are common experiences for most of us. When we find controls that are too hard to work, we either give up, reduce our attempts to use a given device, or must relinquish some of our independence and wait to find someone else with greater strength to activate these overly tenacious switches or controls for us.

The opposite, of course, can be true as well with switches and controls; some are so easy to move (so sensitive) that they can cause device operation problems. The selector lever on your car that shifts into reverse with the slightest bump or the brake on a wheelchair that releases easily and unexpectedly (especially on a steep slope) also come to mind. In an ideal sense, each switch or control will have sensitivity characteristics that are matched with the control site(s) and the physical strength of the user(s) most likely to operate them, as well as with the severity of consequences of their activation.

In consumer products, this appropriate level of sensitivity for optimal user convenience and safety is typically configured for midrange users among the normal population—people of average size and average strength. In many AT applications, we probably find a greater range of variability of muscle force capabilities for switch and control activation among potential users, and must therefore select switch and control items that have the capability for a greater adjustable range of activation-force requirements. Greater or lesser sensitivity for activation can become an essential human factor for successful, reliable, safe operation of AT devices.

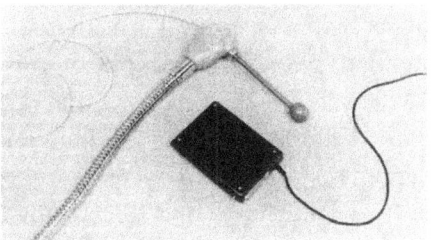

Some switches and controls are more sensitive than others, requiring less muscle force to operate. The reverse can also be true. In this photo, the wobble switch (top) is more sensitive than the treadle switch. Can you identify differing sensitivity among switches you encounter?

In some cases in AT we are working with persons who may have limited ability to generate sufficient muscle force to activate many controls or switches. These conditions may include any of the neuromuscular diseases, CVA, amputations and surgeries, and other conditions that weaken the patient's motoric abilities to transfer force into a switch or control. Occupational and physical therapists, rehabilitation engineers, and physiatrists, among others, may empirically measure the amount of muscle force an individual can generate with any given control site in different positions. By quantifying the amount of pressure or deflection a person can generate with movements of fingers, hand, foot, head, knee, or other control sites, these clinicians can quantify in grams per square centimeter the muscle force their clients can produce, and then prescribe switches and controls that match these muscle- force capabilities. Switches and controls can be designed and custom made to meet the sensitivity needs of a given client. Also, commercial switches that have adjustment capabilities can be prescribed, secured, and individually configured to meet more precisely the special needs and limitations presented by a given client. To our knowledge, no universal industry standards exist across AT devices for what is optimal or even typical sensitivity of component switches and controls. Each manufacturer seems to design and construct a best guess or one size fits all approach into these items. Commonly, they function well for most tech users.

Occasionally, a device appears on the market in which the design and construction of its switch or control sensitivity does not match well with its typical intended users. An example of this was the popular, early version of Muppet Learning Keys, an expanded-membrane keyboard. This adapted input device had laudable intent: It offered a fun, colorful, engaging first keyboard focused on Muppet characters and their activities. It was a valuable aid to learning adapted computer access and a functional motivator for early literacy abilities and computer access skills. In actual practice, however, the smooth, circular, pressure-switch activation sites (keys) on its surface were quite hard to press. In fact, they required so much muscle force to press them that many of the normally developing 4- and 5-year-olds, and all the physically disabled young children with whom we have used the device, have had to have adult assistance in pushing the keys they wished to activate. The intent of the device was excellent: an attractive, motivating, adapted keyboard to help young children and children with disabilities learn to access computers. In actual practice, however, the keyboard was often mismatched regarding sensitivity of the keyboard pressure switches that few small or disabled individuals could operate it with their own muscle force generated by arms, hands, fingers, mouth stick or other control sites.

Quantification of muscle force, as we have mentioned, is a desirable goal. In an ideal world, we could purchase switches and controls that would match the specifically prescribed, constant sensitivity needs of our clients' control sites in all positions and at all times of the day under all levels of fatigue. Such easy universal quantification capability does not exist for most teachers, clinicians, parents, or family members. Perhaps the best practical solutions regarding sensitivity lie in recognizing that this essential human factor can be critical in AT access and attempting to try out and interchange a variety of commercially available switches and controls, as available, that may meet the varying sensitivity needs of our clients. This approach is the usual pragmatic, trial-and-error method that most practicing teachers and clinicians use in considering sensitivity, but now with an eye to this particular human factor.

A simple, applied method of measuring switch sensitivity that we use may be of value in helping you to quantify and report sensitivity measures for various switches and controls. This method consists of securing a roll or two ($2.00 each) of U.S. nickels (5-cent coins) at a bank or business. With the weight of each nickel established at almost exactly 5 grams (5.0582, to be exact), a series of nickels can then be stacked on any given keyboard key, switch, control, or other switch or control activation devices. (Pennies since 1980, incidentally, weigh almost exactly 2.5 grams each.) If the nickels are stacked carefully, a relatively accurate, replicable method of measurement of switch sensitivity can be produced. The resultant pressure data can be used as a practical folk method to compare, select, and configure switches and controls to match the sensitivity characteristics of clients, the strength of their various usable control sites, and the various positions of control site use they may employ throughout the day in various activities from various positions. Using the nickel method, the sensitivity of a switch can be quantified as an "X nickel" (or gram) switch, and thus reported and compared to other switches or controls that may be considered for a client or student.

This method of quantification is simple, inexpensive, and surprisingly meaningful, especially when sensitivity data must be conveyed from one teacher or clinician to another via written reports, individual education plans, and other formal written documents that preserve critical information about a student or client in situations where personal contact and discussion among subsequent teachers or clinicians may not be possible. The nickel method, as simple and perhaps absurd as it sounds, yields replicable, valid sensitivity data—and is much cheaper, plus more universally available and interpretable by others than the use of the more professional engineers' strain gauge designed for formal measurement of sensitivity.

A similar effect can be accomplished by use of an inverted hanging hook postal or fishing scale calibrated in ounces. A pencil can simply be put eraser end into the hook of the upside-down scale, then the pencil point pushed down on to the switch or control to be measured. The ounce level reading is simply read off the sliding analog scale at the precise moment of switch activation. Taking three to five readings is recommended to enhance accuracy and reliability.

An advantage of this method over the nickel method is that different parts of a larger plate or rocker switch can be measured more specifically and precisely. Results from this method of sensitivity quantification can be similarly valuable to, though perhaps more fleeting than, the nickel method. Both are practical, simple ways for teachers, clinicians, and families to gain useful, replicable data on switch and control sensitivity.

Selectivity

Selectivity refers to the likelihood that technology users will activate the switch or control they intend, not one close to it. In other words, how selective are the switches, controls, or other interface activation sites needed to communicate with a device? How easy or hard is it for a person intending to use a given activation device to touch, pull, turn, or press the desired switch or control and not one that is either physically near it—or one that looks or feels similar? Selectivity relates closely to the size of the specific control site a user is relying on (fingertips, palm, ball of foot, back of head, etc.), and to the proximity of other switches and controls to the one(s) desired for activation. The closer together that switches or controls are, and the larger the relative surface area of the control site, the more likelihood that misactivations of neighboring switches and controls will occur. Quite simply, for example, it is easier to hit the wrong keys on your computer keyboard if you are using your heels to type than if you are using your fingertips; the keys are too close for the former control sites but are designed and spaced just about right for the latter. Throughout general consumer technology, most of us have probably attempted to interact with devices that have poor selectivity characteristics in their controls. For instance, the tiny "buttons" on some sports watches and pocket calculators come to mind as items with which most of us have had experience.

Imagine now how physical, motor, and sensory difficulties may further affect a user's access to these or other switches and controls that are not matched to their control site characteristics and special selectivity needs. In many instances, proper selectivity is compromised for general user convenience or design appeal if miss activations due to physical spillover of attempted activation movements do not cause serious consequences.

The sports watch example was a case of a common technology wherein no serious harm is done if the wrong switch is pressed because the buttons are too close together. The same may be true of a car radio's selector switches and other controls, although the distraction and repeated, attention-diverting attempts at corrections created by a driver's miss-hits while driving may potentially lead to serious safety consequences. On the other hand, envision an AT user with severe athetoid cerebral palsy who is trying to manually select important items for urgent communication on her AAC device. She may be able to see the desired activation sites and be able to target them accurately and promptly with her left index finger control site when she is not rushed or under time pressure. She also may be able to generate the muscle force needed to activate the pressure switches underlying the array of vocabulary items on her VOCA where switch sensitivity is matched with the muscle force she can generate. But now comes a situation where her communication is urgent, and she misses the intended activation sites on her device over and over. She keeps hitting neighboring items that are not what she intends. Her repeated struggle to locate and activate the desired items that she urgently wants to communicate is, in this case, hindered by the selectivity characteristics of the VOCA. The items on the array are configured too close together for her motoric needs, causing her to miss hit onto the other items surrounding the targets she is attempting to activate. In addition to being frustrating and adding physical and time load to her use of this AAC device, this situation also could be potentially serious because of her reduced efficiency in communicating important, urgent information.

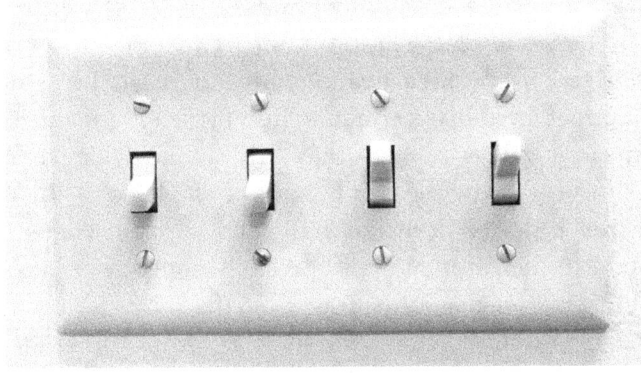

Selectivity can vary across different types and arrays of switches and controls. It is often related to space between activation sites. Sensitivity of switches and controls must also be adjusted as selectivity is varied to minimize error activations

Selectivity in this case could be enhanced by experimenting with how close she can accurately press different activation sites under different stress conditions, and then spacing the items on the array further apart so that her miss hits would be minimized. On some devices, this separating of items to increase selectivity will result in a loss of potentially useful memory space on the surface array of the board—a trade-off for increased selectivity. With some devices, the items and activation sites can be made more selective by spreading them apart without losing memory and icon storage capability on the device. On certain other devices, selectivity can further be enhanced by utilization of features such as dwell and averaging. Dwell is a feature that can be built into the logic of a microprocessor-driven device so that it will accept as a valid activation only a press or "hit" that remains on a given target area for a certain, adjustable length of time. All other hits to other activation sites, if they are not held long enough, are rejected. Similarly, averaging can be part of the logic of the microprocessor-controlled device. It averages the activations to a given site and those around it, then instantly and precisely computes a modal representation of which activation site was hit the most during a specified (and adjustable) period. Whichever site was hit most is computed and selected by the device, again enhancing users' overall likelihood of activating the actual site they intend—that is, improving overall selectivity.

100 words about *MAGNETIC CLASPS AND CLOSURES*

Buttons, snaps, toggles, and zippers can present barriers for people with reduced finger and hand size, mobility, sensitivity, strength, or numbers. Often out of view for the user, closures involve complex, precise movements and adjustments that can be difficult. Elderly persons and users with weak or small hands, such as those living with dwarfism, can find magnetic fasteners empowering. Small magnets and metal anchors with appropriate cosmesis are attached to the stable portions of a garment and to the closing flap or band of the bra, blouse, shirt, jacket, or belt. When near, the parts grip, giving important, subtle tactile and auditory feedback to the user as they offer viable alternatives to other fasteners. See https://www.apexmagnets.com/news-how-tos/magnets-helping-people-disabilities https://www.pinterest.com/pin/90423904994424501

Resolution

Another seemingly obvious yet critically important human factor surrounding switches and controls is resolution. In assistive technology, resolution refers to the process or capability of devices to make distinguishable closely adjacent individual representations of information; the ability of visual (or conceivably tactile) arrays of activation sites to display and convey discrete information in close juxtaposition with other sites or items of information (Kantowitz & Sorkin, 1983). In practical AT terms, this translates as the physical size or, more precisely, the amount of activation surface area that a switch or mechanical control has available to the user to allow for accurate, precise activations. A switch or control device that has greater activation surface area that can be touched and manipulated (generally, a bigger switch) is said to have greater resolution; a switch or control that is smaller in size and surface area would be said to have less resolution. In general, the greater the resolution of switch or control, the greater chance that the user will accurately target and activate it with a given control site; the reverse also tends to be true in most instances.

Several concrete examples of the concept of resolution in general consumer technology are as close as your computer or typewriter keyboard. Notice the space bar and the enter (or return key). They are large in surface area because they are hit frequently and quickly during typing and because they provide critical, repeated functions to the process of entering text to the device. By increasing the resolution of these keys, the designers of the keyboards have increased the likelihood that we will accurately locate and press them with our thumbs and various fingers, especially when we are keying information rapidly.

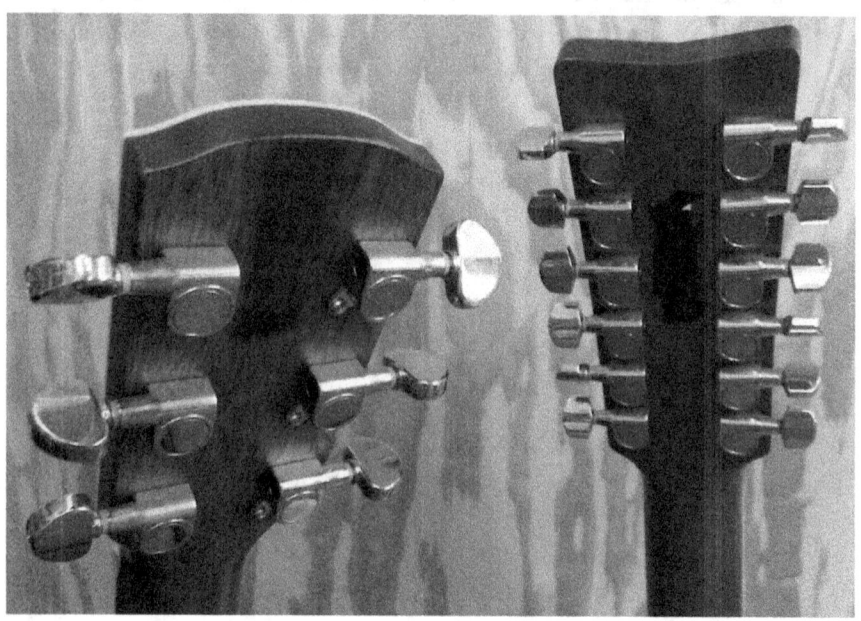

SWITCHES AND CONTROLS: YOUR PASS KEYS TO TECHN... | 153

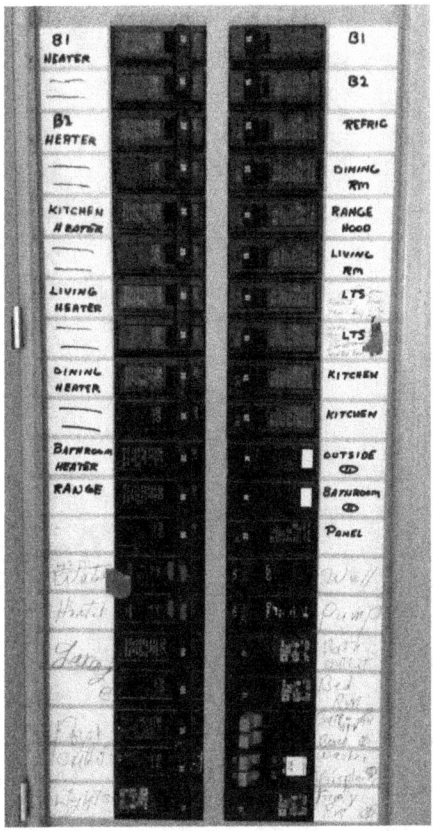

Resolution of activation sites varies with different keyboards, switches, or controls that offer varying sizes of touch targets to suit diverse users.

In AT applications, resolution may be even more important because users commonly have special visual or motor limitations that can cause difficulties in finding and targeting activation sites. By making switch or control activation sites larger and more distinct (i.e., by increasing resolution), AT designers help the users locate and access the information and activation areas they need to control to operate their AT devices. For this reason, the joystick control lever on motorized wheelchairs can be topped with a large, visible round knob or T-grip handle, thus enhancing the user's ability to find it quickly and move it accurately, or early learners who are playing with adapted battery- operated toys may more easily locate and access their toy switches if these are large and visible. Excellent examples of these types of switches with increased resolution are the Big Red and other switches made by AbleNET, and the variety of attractive moderate-sized button switches made by TASH, Inc., and by Toys for Special Children. For this reason, many static-display AAC devices have interchangeable keyguards and overlays, allowing the resolution of vocabulary activation sites to be increased or reduced by changing among 2-, 4-, 8-, 16-, and 32-item arrays. Dynamic display devices also have the capability for modification of activation site sizes. The smaller the number of vocabulary items in an array, the potentially larger each individual activation site can be (approximately 2x2 inches or larger on some devices). As more items are added to a display of vocabulary items on many dedicated communication devices, the more space they take, and the more resolution of each individual item is reduced. More information stored in tighter, smaller spaces increases device memory and information storage capabilities, but it also reduces resolution and thus may adversely affect device access.

Resolution, like selectivity, is closely related to type and size of control sites that the user will be relying on for operation of general technology and AT. Resolution of controls or switch surfaces that are intended to be operated by fingertips should be quite different from resolution of activation sites intended to be operated with an entire hand grasp, foot, or head. In general, the larger the control site surface area, the greater the resolution must be of switches and controls intended for operation with that control site. This seems to be such a basic principle in our general use of technology that we often overlook it. Most of our consumer products are designed to be manual (hands and fingers) direct- select operation, and resolution of control surfaces is matched to the general control site sizes, strengths, and surface areas of the general population. Most users of general consumer products can adapt their skills with control sites to fit the resolution characteristics of the devices (tablets, car radios, watches, keyboards, etc.) they are using. AT users are commonly not as adaptable in skills, so we must configure devices and their activating switches and controls to match much more carefully with users' needs. Matching resolution of switch and control surface areas to the control sites of the AT user is one important component of this individualization process.

Contrast

Contrast of switches and controls from each other and from their surrounding background can be an important factor related to selectivity and resolution. It is also a factor often taken for granted by users of general consumer technologies, and in the design, selection, and implementation of assistive technology for persons with special needs. Contrast refers to the visual, tactile, and/or auditory dissimilarity of adjacent controls and switches. Contrast allows physical activation sites of switches and controls, plus their related visible, palpable, or auditory signals, to be distinguished by the AT user from the background visual, tactile, or auditory fields. Contrast also allows switch and control selection possibilities to be distinguished from one another, particularly when activation sites are in close physical or temporal proximity.

To you, visual contrast appears greater for which symbol? Why?

Visual contrast is enhanced when the color and hue of the activation site is distinctly separate from that of neighboring switches or controls and from the visual background. For example, brightly colored symbols on an AAC device tend to contrast well with a plain, monocolor (black, white, beige, etc.) background. Or on a smaller scale, a figure in a 1 X 1 inch vocabulary symbol, as exhibited by many manufacturers, that is colored bright fluorescent green will contrast well, for many users, with an immediately surrounding background around the icon of brilliant yellow. The utility and popularity of highlighting pens in various colors is based on this principle. For children and adults who have low vision abilities, contrast of switches, controls, and their related activation sites can be enhanced through simple experimentation with opposing colors and hues by their families, teachers, and clinicians. Black on white or the reverse, as well as bright pinks, greens, yellows, reds, and blues on contrasting, more neutral backgrounds, can help make vibrant and bring to life individual items or entire groupings of related items in an array of switches or controls. Creative applications of color, hue, and contrast are remarkably simple and inexpensive enhancements that can be accomplished through use of different colored papers, markers, and switch covers; yet they can add considerably to students' or clients' abilities to see, locate, remember, and accurately access target activation sites to be selected.

Similarly, tactile markers and landmarks that differ from background textures can be added to individual switches, controls, and larger arrays of these devices to aid in locating and selecting intended items. Many standard computer keyboards, automobile controls, and telephone keypads, for example, use tactile markers to assist use by the general consumer population. Consider how much more important these landmarks may become when the users of general or assistive technologies have special sensory, cognitive, or physicals needs.

Tactile contrast can add sensory data essential to discrimination and memory of activation sites, especially those in close physical proximity. Materials with different textures can be taped or glued on to the surfaces of adapted play switches, AAC devices, EC units, wall-mounted light switches, home appliance controls, and so forth. Common materials that work well include small pieces of different textured sandpaper, Astroturf indoor-outdoor carpet, various fabrics with different feels, plastic packing peanuts, bubble wrap, and others readily available around most homes, schools, or offices. Materials used for enhancing tactile contrast on VOCAs and other display panels should be soft and safe for users' motoric and control-site characteristics but should also be durable enough to be fastened and remain in position on devices, switches, and controls. Special care must be given to select materials and fasten them well so that developmentally young users will not be able to put them in their mouths or be harmed by the tactile contrast markers if they attempt to mouth and/or swallow them. Use of contact cement or other strong, super adhesives can help prevent this possibility.

Auditory contrast (different sounding tones and signals) can be particularly important when stimuli are presented in close temporal juxtaposition. The variety of signaling tones used in telephone communication (dial tone, busy, phone left on or off hook, etc.) or in many standard computer functions (sign-on sequence, dial-in sequence for e-mail, end of page, etc.) are examples that we encounter in general consumer use of technologies every day. These contrasting auditory stimuli usually can be altered to suit user tastes and special needs in many standard, off-the-shelf software control panel settings for computers. To some extent they also can be customized with telephones and with other equipment wherein distinguishing of auditory signals is important (front or back doorbells, alarms of various types, sirens, etc.). Auditory contrast can be established overall by variation of three major parameters: frequency, intensity, and quality or timbre of sounds presented. Time duration of sound presentation, including pulsing, sliding, or wailing of any of these parameters can add auditory contrast between information-bearing foreground sounds and other background noise.

Whether visual, tactile, or auditory, the primary concept in enhancing contrast is to make the designated items in the foreground that must attract our attention stand out from others. Our goal is to make these switch and control items in an array sensorily distinct and readily distinguishable from neighboring items that could also be selected, and to make each selection possibility similarly distinguishable from its surrounding larger visual, tactile, and/or auditory background. Creative experimentation with the use of light, color, hue, textures, and sound with our adult and child AT users to address their individual needs can help accomplish meaningful contrast of switches and controls from one another and from their background.

Feedback

Feedback from switches and controls is the essential communication link that tells us, the users, what it is we are doing with the device and how it is responding to our control efforts. Feedback from switches and controls may take many forms, including tactile, kinesthetic, proprioceptive, visual, or auditory, plus combinations of all these modalities.

The centrality of feedback to human performance with tools and devices cannot be overstated. Skilled control or operation of virtually any device, no matter how simple, is nearly impossible without meaningful regular feedback of information from the device to the user.

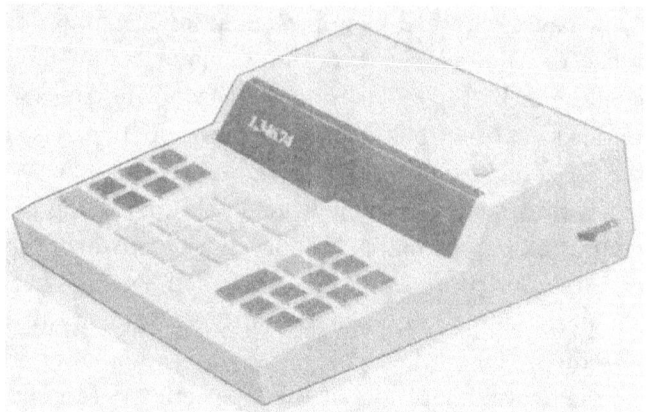

Visual feedback on this desktop calculator is provided by an LCD or LED readout of characters entered, plus a simple glowing ON indicator light.

Even a task as simple as writing a sentence with pen on paper becomes difficult and vastly reduced in precision if you close your eyes so that you cannot monitor visually what your pen and hand are doing. Try it. You will be able to write, of course, but your handwriting will not be as fast, accurate, or clear as when you get constant visual feedback about pen and paper with your eyes open. As another simple experiment with feedback, make a tape recording of yourself playing your guitar or piano softly—while you have ear plugs placed firmly in your ears. When you play back the tape recording and listen to it, you will probably discover that the temporary, artificial reduction in your auditory monitoring skills significantly affected your skill in playing.

Some thought experiments in the importance of feedback that we do not recommend that you try are skating or driving with your eyes closed, singing a solo in church while your personal headphone stereo set pumps loud rock music to your ears, or covering over all the keys on your computer keyboard while you write that important research paper that is due tomorrow. The results of all of these are likely to be disastrous, even fatal. Yet the critical nature of prompt, accurate feedback to our actions in any of these situations relating to operation of technology is underscored by each example.

Latching and Variability

Other important switch and control characteristics include latching and variability. Latching is the capability to stay in a steady-state position (on or off, for example) that can be built into a switch or control. The most common example of switches with binary latching that we all encounter each day are wall-mounted light switches. We turn them on when we enter the room, and they stay on. We turn them off when we leave the room, and they stay off. Seemingly simple factors, yet critically important to our use of that particular technology. It would be inconvenient to have to hold pressure on a switch the entire time we wished to be in each lighted area.

Whereas latching involves essentially discrete (binary or other) state conditions for a switch or control, variability means that the interface may be adjusted for various states of intensity, frequency, or rate. Examples of switches or controls with variability include the dimmer switch on your dining room wall, the volume control on your radio, and the accelerator control on your car. Variability in a switch or control presumes the ability of the user to reduce the quantity being controlled to zero, and then to adjust the control device up and down to various higher or lower levels of output, depending on the application.

In many AT applications, latching is found infrequently except in environmental control units. The AT areas of AAC, adapted play, adapted computer access, and mobility applications all rely more on a steady state of off via switches and controls that require constant activation, ones that must be held for as long as the user wishes the technology to operate, rather than interfaces that allow latching.

Variability in switches and controls is particularly critical in mobility applications, where rate of acceleration, speed, rate and angle of direction change, and braking must be anything but on-off in nature. Just as in driving a car, powered assistive mobility relies on incremental, highly controllable functions that must be capable of being initiated, modified, and terminated at the will of the user at any time. Latching in such cases must be used cautiously and is typically found only in powered wheelchairs configured for skilled users. The user has the option of latching the joystick control into a straight-ahead running mode, appropriate for long stretches of sidewalk or road, where turns or stops are unlikely to occur. Such latching capability, in this case, though potentially dangerous for those with reduced motor, sensory, or reaction-time capabilities, does reduce physical effort. Users can latch the forward motion control into a desired speed and then remove their hand or other control site from the direction/acceleration control to rest. Latching and variability in switches and controls both have their advantages and dangers, and must be used depending on the needs, limitations, and characteristics of the student, client, patient, and family who are relying on assistive technology.

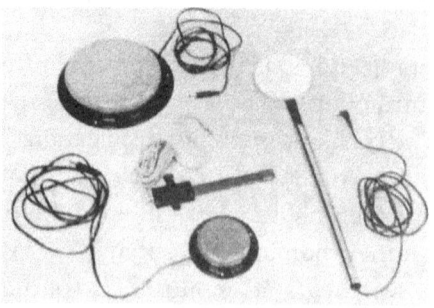

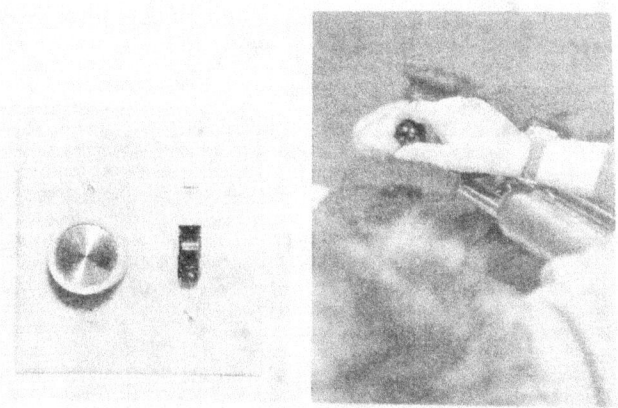

Dimmers, toggle and adapted switches, plus powered wheelchair joysticks exhibit and feature various degrees of latching and variability.

Positioning and Mounting

Positioning and mounting of switches and controls can be critically important to their efficacy for any user of enabling devices. A switch or control that is not placed within optimal reach for whatever control sites a person may be using can cause undue fatigue, reduced accuracy and response time, and overall reduced efficiency. This is true for any user attempting to interact with controls, switches, or other interfaces of a consumer or AT device. It becomes especially important for persons who have movement limitations or limb deficiencies; a switch or control that is out of reach and therefore inaccessible may as well not exist.

Positioning and mounting of switches, controls, and devices is most properly the province of the occupational and physical therapist and other specialists including rehabilitation engineers and technicians, physiatrists, and others who treat, teach, and collaborate with the AT user and his or her family. As a collaborative team, these professionals and others can address specific switch and control positioning and mounting issues for a given client based on the client's control-site needs and characteristics. They can engage in an ongoing problem-solving venture over time to best meet their client's changing switch and control positioning and mounting needs.

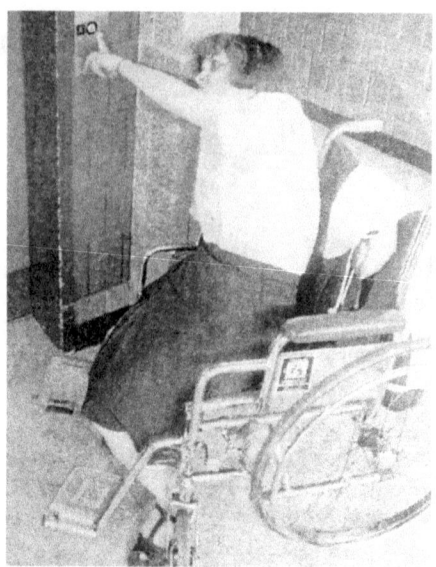

Switches and controls for mobility and environmental control that are positioned out of reach are of little use to consumers who have special needs not addressed by designs of essential technologies such as an elevator.

Although it would be convenient if this addressing of positioning and mounting needs were a one-time event, this area of AT practice involves an ongoing process that should be constantly in a state of revision and updating as needed. Continued attention to these needs for our students, clients, and patients is essential for several reasons:

1. A person's control-site characteristics can change over time: Range of motion, strength, speed and accuracy of targeting, and the like may improve or worsen depending on the nature of the disability, limitations, or illness.
2. The body parts that a person uses as control sites may change because of skill building or loss, tissue deterioration, or other factors related to the individual's health and tech needs across different settings (hospital room, home, school, driving, employment, etc.).
3. The assistive technology itself may need to be accessed in different or multiple ways depending on how and where it is

used, improvements in devices, or need to adapt use of devices to various applied environments (all per King, Schomisch, & King, 1996.)

Although it would seem obvious that switches and controls placed out of easy reach would reduce anyone's ability to interact with technology if we are not the user—and if users cannot clearly express their access needs—it becomes more difficult to determine just where and how these items should be mounted and positioned. Careful observation and data taking are needed as various configurations are tried. Close, continued professional consultation with users, families, and other care providers on the specifics of switch and control positioning and mounting over time and changing needs is essential.

Composition and Construction

Composition and construction of switches and controls refers to what these interface devices are made of and how well they are made for actual use in assistive technology by real human users. In general, the switches and controls that are now commercially available for AT applications are well designed and field-tested in a variety of trying conditions. They are composed of durable, appropriate materials for safe, reliable operation in a multitude of demanding settings. Some important innovators and suppliers of switches and controls for AT include the following companies: AbleNET, Crestwood Communication, Prentke-Romich Company, Technical Aids and Systems for the Handicapped (TASH), Toys for Special Children, and Zygo Industries. Further details on specific contact information for these and other AT-related companies may be found via online searching, plus TRACE and RESNA data bases, among others, as listed in the Resources and References appendix at the end of this book.

Composition and construction of switches and controls include the materials from which the external housings and contact surfaces are made. They also include the quality and type of electrical contacts and internal wiring in the device, and the external wiring, leads, and connector plugs or jacks that allow the switch or control to be connected with AT devices or other switches and controls. Most AT switches and controls are now made of a combination of molded plastics, plus steel and/or aluminum housings and parts, with spring steel and copper electrical contacts. Their connecting leads are commonly paired, stranded number 22 (or similar) gauge hookup wire, terminating in either a ¼ or 1/8-inch (3.5-mm) phone plug. They are rated for use with 1 to 12 volts DC (direct current electromotive potential) and for use with very low amperage electron flow (in the 0.5 to 1.0 milliamp range maximum). AT switch and control units readily available on the market are generally sealed, smooth-cornered, and electrically and mechanically tight in construction. They are resistant to moisture from drooling and immersion and to mechanical damage; they resist electrical shorting or physical splintering and other breakage that could result in possible tissue trauma or other injury to the user, especially during rugged use.

It is paramount in AT that any switch or control which is used in conjunction with a special-needs student, client, or patient must be able to withstand safely the potential for electrical shorting as well as for corrosion of materials during use. These can be caused through repeated exposure to the user's drooling of saliva, spills of food and liquids during eating and drinking, hard- contact physical use and rough handling resulting from a user's rigid, spastic control sites, and from his or her enthusiasm and aggression. It is also common for switch-operated devices to be used with special-needs persons in the bathroom, including in the tub or shower, as well as in wet or sandy play areas, wading pools, swimming pools, and at the beach with fresh or salt water. Switches must be safe in these settings, too.

A few words about electrical safety are appropriate at this point. As mentioned, part of the composition, construction, and durability of a switch or control must be measured by its electrical safety. Switch or control contacts that become broken or exposed can pose a true shock, bum, or even electrocution hazard to the user; so can broken, exposed wiring or contacts in plugs, adapters, or line cord wiring that accompanies the switch or control. In general, most AT devices will present voltages to the user's switches or controls that are 12 volts direct current (DC) or less. This is well within a typically safe range because voltage ratings below 30 are commonly considered safe for most adults. However, the other aspect of electron flow, amperage (current of flow), is also a major consideration. Most switch-operated AT devices, as stated, will have amperage levels of 0.5 to 1.0 milliamps or less present in the switches or controls available to the user. If a break in electrical insulation in the switch, cord, or plug should occur, these low levels of current may be just perceptible as a small tingle or shock or may not be perceptible at all to most users. Just how these low amperages and voltages might affect a small child who is drooling or who also may be seizure prone (or has other neuromuscular special needs) is not known. For most users, amperages under 1 milliamp are just perceptible, with the range for harmless current extending up to 5 milliamps (Wolfgang, Kearman, & Kleinman, 1993). Current flow of 10 to *20* milliamps is the lower level that might initiate sustained muscle contractions, with 30 to 50 milliamps the levels where pain, possible fainting, or not being able to let go of the wire occurs. Amperages in the 100- to 300-milliamp range disrupt heart rhythm and may electrocute some persons. At levels of *6* amperes and above, a person will experience sustained heart contractions, and burns are likely (as adapted from Wolfgang, Kearman, & Kleinman, 1993). Suffice it to say that electrical and mechanical integrity of switches and controls, and appropriate levels of operating voltage and amperage, can be extremely important factors in their selection and use in AT, especially for special-needs users who may present unique health and behavioral challenges.

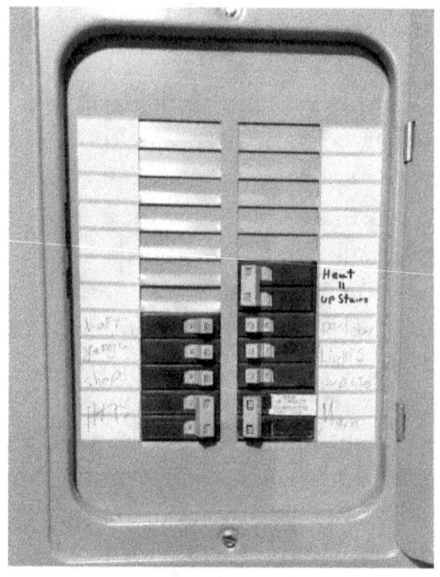

Rugged, sealed construction is essential in switches, controls, adapted keyboards and other consumer and assistive technologies.

In addition to the reality of these environmental hazards, many children and adults, particularly in rural settings, must spend considerable time outside in high ambient humidity, rain, snow, or other precipitation while they are waiting for transportation to school or work. They also may spend considerable time outside in the elements while moving between bus and home, or car and garage, or just playing, recreating, and socializing out in their home yard or school grounds. Any switch, control, or interface device that becomes such an integral part of a person's life for mobility, communication, environmental control, play, and other areas of critical daily-life needs, must be able to withstand the rigors of repeated, sometimes long-term exposure to the environment. They must be able to function dependably through moisture, temperature, and mechanical stress placed on these devices by constant vibration during movement, contact with hard structures in the environment (door frames, trees, walls, etc.), bumps from other children and their bats, balls, or swings; and rough, hard "hits" by the user and others who may not always activate the switch or control with the greatest of care or gentleness. Many users of consumer as well as assistive technologies can be demanding and "tough" on their switches, hitting them hard, pounding on them, and running them into obstacles in the environment. These "levers" with which our AT users move their worlds must be able to withstand all these rigors. Manufacture and design, as well as our selection, purchasing, prescription, and use of switches and controls for AT must be guided by knowledge of how extreme demands on their component material types and construction integrity can be.

Another area of switch and control design and construction that is critical for users in colder areas of the world is low (sometimes very low: -50 to 0° F in many places) ambient temperatures. Plastic parts of switches and controls can become brittle and break or even shatter when pushed, pressed, pulled, or turned in subzero cold. These temperatures are experienced by many children and adult AT users, as well as their devices, while they may be outside waiting for transportation or playing in the northern United States and Canada, as well as in other countries in the temperate and near polar regions of the planet. The obvious danger to users from pointed, jagged, broken plastic parts, as well as the inconvenience from this potential response of plastic switch and control materials to extremely cold weather, is not uncommon and could have potentially serious ramifications for the device and for the user's safety. Care must be taken to limit exposure of switches and controls to extremely cold weather and to construct these interface devices from materials that can withstand the widest possible range of temperatures without breaking down or shattering.

Overall durability under all conditions is, of course, a factor in selection and use of switches and controls. The ability of all devices used in AT to withstand contaminants found in the real world, such as water, sand, and salt, temperature extremes (heat and cold), and direct sunlight, without breaking down and becoming brittle, shattered, and dangerous is something that manufacturers must consider in design and construction. Consumers of these products should take into consideration the long-term durability of interface devices as we recommend them for our customers, students, clients, and patients. Research and detailing of clinical experiences with materials in these conditions are needed.

Closely related to durability of switches and controls is their ability to be repaired or replaced as necessary when difficulties do arise. In general, suppliers mentioned earlier in the text and most others in the fields of AT will warranty and replace a switch or control that fails under normal-use conditions. The problem can be that normal use can be difficult to define for many AT users; their switches and controls can be put into some highly demanding situations, as we have already discussed. It would be difficult for any device to continue to function well when pressed to the limits that some AT switches and controls are over time. Nonetheless, most such devices work well and reliably in the presence of wet, cold or hot, sandy, and other conditions.

When a switch or control does break, it is sometimes possible for a knowledgeable teacher, clinician, or family member to use a screwdriver to open and clean out or make simple repairs on basic external switches or controls for toys, AAC devices, and adapted computer peripheral controls—if dangerous electrical currents (amperages over 5 milliamps) are not present. All such AT devices should be turned off, unplugged, and the switch or control totally disconnected from the device before any repairs are attempted. Better yet, if in doubt, send it out—to the supplier to be fixed.

It has been an informal rule among tech folks that 90 percent of problems with electrical equipment is in the connectors, so the circuit continuity should always be checked in plugs, jacks, wires, and switch contacts before one decides that a switch, control, or an entire AT device is defective. Often a few simple tightening, scraping, or resoldering maneuvers on switch and control contacts, plugs, and jacks can restore the interface device to full function by removing corrosion and improving electrical conductivity. If a switch or control cannot be repaired by the user and those working on his or her behalf in the school or clinic, the item can be sent back to the manufacturer in most cases for repair at nominal or no cost other than postage. Replacement of switches and controls that have been used a long time and that have become severely worn through use and weather, or corroded by food, saliva, toothpaste, bathwater, and other contaminants from the user's daily environment may occasionally be necessary. By dealing with the large, reputable suppliers in the AT field, one is virtually assured of a dependable, stable supply of replacement devices of the same or even better quality through the years. Although it is always wise to budget ahead for switch and control updating and replacement, the costs of these interface items in AT is typically not excessive. Most of these items, fortunately, are in the tens of dollars (U.S.) range, not hundreds or thousands of dollars as are the larger pieces of AT gear that they help us to control and communicate with. It is wise to get on the mailing lists of the major switch and control supply companies in AT so that one can stay abreast of new items, costs, and other innovations as they occur.

Switches and controls take a lot of wear and tear and, frankly, abuse. Being able to repair or to replace them quickly and appropriately can create much needed independence for users, and their teachers, clinicians, and family in any given locale by letting them repair and replace interface items on their own as needed. This helps to restore AT function more promptly for users and prevents unnecessary system down time. From an engineering perspective, a system is only as strong as its weakest link. For many other AT devices, switches and controls are often that weakest or at least most used and most damaged link. Being able carefully and wisely to select, prescribe, purchase, use, and repair or replace them at a local-user level can help keep AT working and beneficial with few gaps in service for those who depend on it.

Cosmesis

Cosmesis, as we first discussed in earlier chapters, refers to the relative attractiveness, or lack thereof, of an item to the user and to others around him or her (Vanderheiden & Lloyd, 1986). This is such an important concept and human factor that we will expand on it further in this section, with special attention to cosmesis of switches and controls. These AT items typically are close to or in contact with a user's body—they invade the user's space—and merit special additional discussion as cosmesis relates to them.

The term cosmesis, with the same root word as cosmetic, has to do with an area that is important to each of us personally once we grow beyond our own early childhood years—namely, how we look to others and to ourselves. Much of the U.S. and world economy is based on clothing, accessories, cosmetics, and other preparations designed to make us appear more attractive to others—or at least to believe that we appear more attractive by using them. Particularly regarding items worn, carried, or mounted close to or on our bodies, we all, in most societies, at least, seem to be quite concerned with how things look with us. How our eyeglasses or contact lenses or jewelry or purse or belt or any number of other apparel, accessory, or jewelry items make us look is a valid concern for almost everyone. We all like to look good and, maybe even more important, we hope to avoid looking out of style, geeky, dorky, or otherwise separate in appearance from our peers and associates.

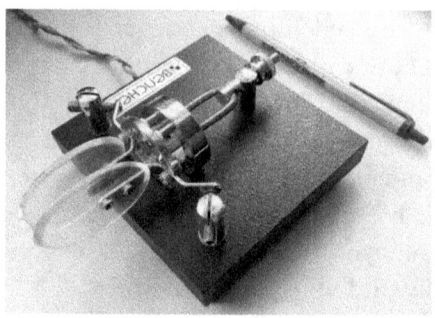

Cosmesis is important to consumers, but is not always considered in the design, selection, and configuration of controls, switches and technologies. Cosmesis, overall appearance, and cultural elements can be critical factors in users' acceptance or rejection of tools and devices.

How we appear to others occupies much of our time, money, and thoughts in this current consumer culture. Why should those who use assistive technology be any different? Should they not also legitimately be concerned about how they look to others and themselves amid the devices, switches, controls, mounts, and other apparatus that accompany their essential AT devices?

Some compelling examples of how crucial cosmesis can be in the lives of children (and probably all of us) derive from research in pediatric cancer. In recent discussions with colleagues regarding elementary school-aged cancer patients, professionals reported how they have asked these children to describe pain and painful experiences. The clinicians hypothesized that the children would respond with comments about needles, surgeries, and transfusions. With rare exception, however, the children talked much more about being bald, being laughed at, and being excluded from their peer group because they looked different. These were the experiences they described as causing them pain. We might believe that in life-threatening cases such as cancer, young patients would come to regard appearance (cosmesis) as a distant concern from the physical pain involved in their struggle and survival. Yet information from interactions like these may indicate that the social pains of rejection, separation from peers, and humiliation based on appearance may be as great or greater than the physical pains related to struggle with a terrible disease.

As we grow through our teenage years into adulthood, if we happen to be users of specialized AT, we may better understand the reasons that we must be configured with special input switches or devices. Nonetheless, our desire to fit in, to believe that our appearance is accepted by others, and to look as good, as possible does not abate. In adulthood, we share the common concerns of attracting and keeping a life partner and of finding and holding gainful, fulfilling employment. Some of the attainment of each of those common life goals for adults in most societies relates to our personal appearance and the acceptability to others of how we look and fit in with the people around us.

Recent accommodations from some companies for switch and control cosmesis have been interesting to watch. The array of attractive, distinct colors available in switches from companies has been growing, and the style and contouring of many of these switches have been perfected to give them a pleasing, rounded form that can be compatible for use and fuller inclusion into life of the AT user. Some even look "pretty cool," according to some children. Although addition of any external switching or control apparatus to a person's immediate body space has an impact on overall appearance, the interface devices that are now available can be quite attractive. Many users report that they are at least tolerable to have in their presence much of each day as necessities for access to their AT devices.

New on the market now too are switch covers, soft fabric covers sewn from attractive contemporary fabrics that can be slipped over standard button and plate switches. They allow users and their families to customize the appearance of switches and controls around the user so they match clothing and circumstances more closely: informal colors and designs for fun, play, and socialization; more formal colors and designs for work, church, and other settings where a person would typically dress up in clothing and accessories. Switch and control coverings in and of themselves can help add to the acceptability and cosmetic appeal of a user's interface devices. They can also add an additional measure of enhanced cosmetic hygiene by being washable and interchangeable. As some AT users may drool, spill food or drink on a control or switch surface, or otherwise soil the control interface area, the removable cloth covering can be laundered while it is replaced by another interchangeable covering.

In our experience, switch covers have worked the best (at least most easily) with the relatively flat button pressure switches that are mounted for head, chin, or hand activation by AT users. These switches include the Big Red and Jellybean switches from AbleNet, the Buddy Buttons from TASH, and similar interface devices made by other commercial switch suppliers. Covers for these switches are simply a sewn envelope of fabric, which can be of various colors and types. The fabric envelope quickly slides over the switch and then fastens shut with a zipper or Velcro-type closure along the open side. Different-sized covers are available to match different-sized switches of varying diameters and thicknesses. When the cover gets dirty, it is simply opened and removed from the switch. A new one is slipped over the switch and fastened on along the closure side, while the soiled one can be washed or even discarded if necessary.

An investment in these covers can truly enhance the cosmesis of an AT user's overall configuration on his or her workstation, wheelchair, and/or bed. A feel of greater dignity, attention to detail, and desire of the user and family to be included in as much of life as possible are conveyed by use of switch covers, in our opinion. They help reduce some of the stark, hard, institutional feel that bare switches and controls can convey and are worth the few dollars of investment for the cosmesis and the personal hygiene benefits they can offer an AT user.

Switch and control covers should, of course, be used only with the interface devices that they can safely cover without restricting or interfering with their operation. If the fabric will cause jamming, binding, or reduced function of the control or switch in any way, the use of a cover is not wise. The construction of switch and control covers, incidentally, makes a satisfying project for family members or others who wish to help a person using AT. Sometimes it is hard to suggest specific tasks for well-meaning persons who offer to help an AT user. Switch cover making and updating in cool, fun, newly designed colors and fabrics can be a relatively simple task that individuals or groups who wish to do something concretely helpful for an AT user and family can contribute.

Sensory Defensiveness of the User

User tactile and sensory defensiveness can be significant factors in switch and control acceptance, use, and efficacy for a given user. Some potential AT users, particularly, it seems, many with other concomitant disorders and special behavioral or cognitive needs, may exhibit mild, moderate, or even extremely adverse avoidance reactions to touching certain surfaces, textures, and temperatures of objects (Burkhart, 1997; Goosens & Crain, 1992). Tactile and sensory defensiveness may often be seen in young children with other special needs. These aversive types of defensiveness and avoidance may manifest particularly in the case of switch or control use learning because they must touch the surfaces of these devices often and under the encouragement or coercion of their parents, family, teachers, clinicians, or others attempting to teach them AT skills. The child or adult who displays tactile or sensory defensiveness exhibits considerable negative response, even catastrophic avoidance behaviors, to touching certain textures, temperatures, and other characteristics of object surfaces.

Often, adverse behavioral responses are triggered in these users by switch, control, toy, or other device activation surfaces that are too cold, rough, hard, smooth, irritating or other extremes of tactile sensations. Many times these concerns, if known beforehand by family or AT professionals, can be addressed by teachers, clinicians, and parents through selection of switch or control types that will fit with users' needs, or through use of switch and control covers made of various materials that can make the device more acceptable to the children or adults who will be using it from the perspective of their tactile and sensory needs.

In addition to modifying the contact surfaces of switches or controls that we expect users to touch, regimens for desensitizing the special-needs switch user to various related sensory demands have been described by Goossens', Berg, Lane, and Crain (1987) and Goossens' and Crain (1992). Each of these programs can be carried out by teachers, clinicians, family members, or other skilled care providers. They include working through a series of incremental steps that involve the AT user making progressively longer and firmer contact with a variety of surface textures and/or temperatures as needed. Users are helped to accept the touch of materials that progressively approximate the sensory characteristics of the switch or control surfaces that they will eventually need to touch on a frequent, repeated basis to operate their AT devices.

The success of these techniques depends to a large extent on the interventionist's skill in choosing sufficiently small, appropriate incremental steps that gradually approach the feel and the temperature characteristics of the target surface that the user will eventually have to touch. Success also depends on moving through this touch and acceptance hierarchy with a given child or adult in a relaxed, slow manner so that the user can be progressively desensitized in ways that evoke minimal (or, ideally, no) adverse behavioral responses. A slow, steady, gentle, but firm approach to tactile desensitization often works well with special-needs persons. The primary concepts are to view the

SWITCHES AND CONTROLS: YOUR PASS KEYS TO TECHN... | 175

tactile needs from the user's perspective and to move through the hierarchy in a steady, non-rushed manner, allowing new learning and security with new textures and temperatures to become comfortable experiences for the child or adult AT user before moving on. Proceeding too quickly can mean failure of the effort—and potential failure of the user to accept and learn even basic control skills with their AT devices, whatever the type.

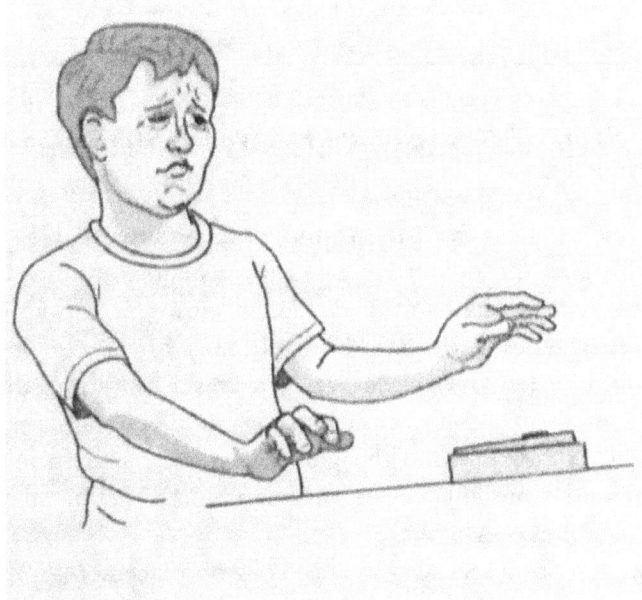

Sensory or tactile defensiveness of an AT user can interfere dramatically with his or her acceptance and successful use of switches, controls, and other devices.

---

100 words about *ADAPTIVE CLOTHING*

Apparel, headwear and footwear for people who are smaller, older, transitioning, or of any age living with challenges or disabilities, combine important, often unrecognized human factors considerations. These include clothing cosmesis and fit for each individual's

body size and characteristics; operation of closure devices such as zippers, button and clasps; compatibility of fabric and materials with special skin or sensory-defensiveness characteristics; range of motion; plus relevant ethnic, cultural and religious considerations. Clothing is protective mobile technology that is also essential to all of us in the form, appearance and confidence it affords. See www.disabled-world.com https://www.mayoclinic.org/diseases-conditions

A BRIEF INTERNATIONAL PERSPECTIVE ON SWITCHES AND CONTROLS

Switches and controls from the primary Canadian and U.S. suppliers listed in this text and others are generally available around the world. With current transshipment capabilities through various carriers, delivery of switches and controls to virtually any place on earth is possible within a few days if desired. Perhaps a more pressing issue is the awareness of teachers, clinicians of all types, families, funding agencies, and others around the globe of products that are available and the essential human factors variations in them that may recommend certain ones over others.

Although consumer switch and control placement for lighting, water control, automobile operation, appliance control, and other daily tasks are reasonably standard in the Western industrialized countries of the so-called First World, this may not be so in Second or Third World, non-industrialized societies. Variations in construction, wiring, plumbing, and appliance manufacture can cause access difficulties, at least temporarily, for some special-needs users. In many developing countries, much of the infrastructure that makes the rest of AT possible is not yet present. Some of our students who have come from rural Central American countries have described how paved roads or sidewalks are not common in their villages, and that electrical supply to homes, allowing plug-in, charging, and operation of electrical AT devices typically is not available at all. Limited access to electrical supply in many countries would, of course, drastically reduce the number of switches and controls available to all technology users, AT or otherwise. Van Tatenhove (1993) mentioned the rarity of electrical supply in rural South Africa and stated that AAC symbol boards there commonly were indeed wooden boards that she would draw on for her students, and that sometimes were subsequently used for cooking fuel in homes. Not everything technological in the world is as it is in North America or Europe. Effective low-tech is still the readily available rule and resource for most people and societies.

The need for expanded international awareness of assistive technology and, of course, its pass keys, switches and controls, is great. Organizations such as the International Society for Augmentative and Alternative Communication (ISAAC) and the Rehabilitation Engineering Society of North America (RESNA), among other groups in Canada, the United Kingdom, the United States, and other countries, play important roles in helping create awareness of, clinical skill with, and availability of AT and related devices on an international basis. Readers are encouraged to learn more about these organizations, and to become active in their international work. Further contact information is listed in the resource appendix of this book. Within the United States and Canada as well as other countries, multicultural issues surrounding AT and people who use it also merit much more study and recognition. With individual reasons and traditions, not all cultures value or regard technology and the interventions and life changes that accompany it in the same ways as mainstream Western culture. Insights into users' family and community structures, and into their cultural uniqueness, can aid in acceptance and use of AT. These insights and sensitivity to users by professionals can assist markedly in the successful integration of AT into the life of consumers, their families, and the community.

---

100 words about *SPEECH AND VOICE ACTIVATION*

Speaking to our technologies, once science fiction, is common reality in homes, cars, work, and entertainment. Advantages of speech and voice input are immediacy and convenience of hands-free tech use.

Disadvantages include smart devices monitoring and acting upon our unintended speech or environmental sounds, resulting in social, communicative or informational tensions. Persons living with inability to produce precise, recognizable speech and voice can be locked out of speech input technologies, as can speakers of languages unrecognized by our devices. See https://slate.com/technology/2018/10/voice   https://www.totalvoicetech.com/five-things-that-can-interfere-with-your-voice-recognition-software

ASSISTIVE TECHNOLOGY AND FITTS' LAW

It would be useful if there were a more empirical method of expressing the lawful relationship between control sites of the technology user's body, and the target activation sites of AT switches and controls. Fortunately, such a description has been devised. Fitts (1954) and Fitts and Peterson (1964) have described a speed-versus-accuracy-of-movement equation that gives us this capability. Several writers have elaborated on Fitts' law, including Kantowitz and Sorkin (1983), Sanders and McCormick (1993), and King, Schomisch, and King (1996).

Fitts' law and its applied extensions to real-life situations have been used for several years in the human factors considerations that underlie design and use of industrial, military, and consumer devices. Most of the switches, controls, and other activation devices that we are accustomed to using have likely been designed with this breakthrough description of the lawful relationship between control site and target as a major guiding force. Based on these authors' and others' work, as well as our own clinical experience, we will speculate on and attempt to extend some of their interpretations of this speed-versus-accuracy trade-off relationship as it may apply to assistive technology.

In its most elegant mathematical form, Fitts' law is expressed as

$$MT = a + b \log_2(2D/W)$$

- MT equals movement time of any given control site (body

part) from initiation of movement to touching the targeted activation site (switch or control surface).
- a and b are empirically derived constants.
- D equals the distance of control site movement from start to center of the targeted activation site.
- W is the width of the targeted switch or control activation site.

Application of Fitts' law to real human-machine interface questions in industry and in machine design has been shown valid and able to predict and describe lawful relationships in human interactions with switches and controls (Kantowitz & Sorkin, 1983). Although understanding and applications of the precise, complex mathematics of Fitts' law are surely important from an engineering perspective, from the viewpoint of most AT users, their teachers, clinicians, families, and care providers, a more generalized, practical "hands on" interpretation is the approach we will use here. Some practical, more clinically relevant extensions of the Fitts' law equation will be discussed, especially as these relate to the perspectives of Baker (1986), Goossens' and Crain (1992), Silverman (1995), and other clinicians and teachers whose work we have read or with whom we have had lengthy discussions and clinical problem-solving sessions over the years. Taking the physics and the science of human factors relationships such as those described by Fitts' law, then combining and interpreting them into useful, usable knowledge and skills for human intervention is constantly the job of skilled teachers and clinicians. We will attempt to synthesize and describe some practical applications of this guiding relationship in control site and switch control use based on study and clinical practice that may be useful for teachers, clinicians, and families of AT users.

This synthesis will be organized primarily according to the components of Baker's basic ergonomic equation regarding motivation versus effort in assistive technology use, as described in previous chapters. To review briefly, the component areas described by Baker include motivation of the user to use AT, as contrasted to the amounts of physical effort, cognitive effort, linguistic effort, and time required for successful technology. In the following discussion, we will speculate on and discuss how essential principles of Fitts' law and its clinical extensions might interrelate with components of the Baker equation and with some of the essential human factors characteristics of switches and controls covered earlier in this chapter.

Physical Effort

Applications of Fitts' law principles to assistive technology may be most evident and valid in physical effort because that is what the equation was primarily developed to describe. Several important aspects can be described relative to our extensions of this equation as it is pertinent in physical effort in AT use.

The first consideration is that, in general, larger (wider) control site surface areas (i.e., fingertips vs. whole palms vs. ball of foot vs. side of knee, etc.) require larger (wider) activation sites on switches and controls for optimal accuracy of targeting (i.e., hitting the switch) at constant speeds of movement. The reverse is also generally true as speeds of control site movement increase. In practical terms, this means that if we or our clients may need to use a control site that is wider than what a certain switch or control is designed for, we or they will have greater difficulty in accessing it rapidly and accurately. For example, a standard computer keyboard has keys with a resolution of about 1 X 1.5 centimeters. These keys are plastic covers over small pressure switches and are designed to be activated with control sites that are about the same width and surface area of our fingertips. If we or our clients could not use fingertips to type on the standard keyboard, but instead tried to use our heels as the control sites, it would quickly become evident that there was a huge mismatch in control site surface area versus activation site (switch or control) area. It would be difficult though certainly possible to operate (type on) such a keyboard with our heels as primary control sites, but our rate (keystrokes per minute) of accurate keystroke entry would decrease markedly, as our error rate of entries would increase because of mistaken activations of neighboring keys.

This comparison of fingertips versus heels makes an interesting experiment in one of the primary, essential applications of Fitts' law. We encourage you to try it with your own computer. You may use a single word, your name and address, or the standard sentence from

SWITCHES AND CONTROLS: YOUR PASS KEYS TO TECHN... | 181

typing class if you wish as your sample text: The quick brown fox jumped over the lazy dog. Be sure to time yourself with both entry methods. Try it with and without correcting your mistakes. You will soon gain some important insights into control site and activation site size matching that you will be able to apply to your students, clients, or patients.

Similarly, most standard toggle light switches on a wall-mounted panel are about 1.5 cm long and 0.5 cm square in width and depth. These switches are also designed to be operated with fingers and fingertips, and work well with our usual matching control sites across human individuals. However, if your control site for operating the lights in your home for some reason of special need became the flat part of your forearm or your elbow area, you will find that it is now going to be much more difficult to put the control site on to the switch you desire to turn on or off—especially if the wall-mounted switches are small or spaced closely. You can learn to do it, but again your speed and overall movement efficiency of activation of the switches will be reduced because of the mismatch of your now nonstandard control site surfaces to the activation site surface. Be sure to try it with various control sites from standing and sitting positions, and then see what happens.

If surface areas of control sites and activation sites are mismatched, rate and accuracy of entries will decrease. Selectivity, resolution, and sensitivity closely interrelate for maximal acceptance and efficiency of switch and control use.

Another excellent example of this aspect of interpretation regarding Fitts' law applied to our everyday uses of common and special assistive technologies is the design of our standard automobile clutch, brake, and accelerator pedals. Control sites for both are typically the ball of the foot or the fuller middle portion of the foot (typically the right foot in the United States, the left foot in many other countries). These typical control sites are intended to match well with the design of most brake and accelerator pedal controls in that these interfaces with the car have width (and overall surface area) that matches approximately the width and breadth of the adult human foot. When you need to shift, stop, or accelerate, you can perform these operations generally, after a bit of practice, without looking at your feet for visual feedback. The activation site surfaces match your control site surfaces closely enough that the chance of your hitting the correct pedal with your foot, even when you may be required to do so very rapidly in heavy traffic or in an emergency stop for instance, is extremely high. If for some reason, the foot-pedal portion of your clutch, brake, or accelerator controls should be missing and only the thin metal support rod of the control is left, you may indeed experience considerable difficulty—and it would take more time—in finding the place to push with your foot to operate the desired stop or go control. This obviously could be a significant inconvenience and safety hazard. In fact, I had a pickup truck in my student days that exemplified these characteristics exactly. Finding and pushing on the protruding brake rod that had no foot pedal portion left on it was a true challenge and fostered my first awareness of Fitts' law in this type of practical mobility application. I do not advise you and your customers or clients to try this experiment.

In each of the cases and simple experiments described here, it is evident that as the activation site area was kept constant and control site area increased, in general, time for successful activation would have to increase to operate that switch or control with increasingly larger area control sites. The reverse, as mentioned, would also be true in general. That is, if you kept your or your client's control sites constant (again, let's say fingertips), but reduced the surface area of the activation sites for switches or controls, speed and accuracy of activation would also likely be reduced—it is harder to find and activate smaller sites. For example, fingers and fingertips work fine in operating the controls of many sport-type wrist watches. The designers of some watches, however, have reduced the activation site surfaces to a size that is so small that most adult human fingers no longer work as effective control sites; the size mismatch now goes in the other direction from what we had discussed earlier. A pen tip or some other small, pointed utensil must be used because it is smaller in surface area and matches better with the activation site provided on the watch. You can probably think of other examples of this direction of control site-activation site match and mismatch regarding practical applications of Fitts' law—from the large laces used to help little fingers tie shoes in kindergarten, to the pressure-switch selector buttons on many soft drink machines, where the widest, largest one represents the product that the company wants to sell the most, and the smaller buttons represent the less-popular items.

You will probably find exceptions to all the foregoing, at least for technology users who have motor and sensory skills within relatively normal range. For instance, it is indeed possible to dial the phone with one's nose, or to turn the TV off by pushing the switch in with one's knee, and so forth. You could argue that these actions can be done quickly, accurately, and relatively efficiently even though the switches or controls described are not expressly designed for the control sites used.

You are right. In fact, these types of creative, divergent thinking exceptions are always possible, and should be recognized. That is one of the main missions of this book. These types of exceptions and the demands they place on us for creative, divergent problem solving become almost the rule for AT users and for their family, teachers, and clinicians, who must constantly learn to adjust nonstandard, even awkward and limiting control site capabilities to the demands presented by a world of direct-select activation-site sizes that most often do not match with their individual needs.

Another aspect of Fitts' law as it relates to physical effort that has practical application in AT surrounds the matter of switch and control selectivity. In general, larger control site surfaces require greater selectivity of activation sites. That is, switches, controls, or other interfaces must be spaced further apart or have other built-in protective features to help users avoid error activations at constant speeds. Again, we return to our standard computer keyboard, and the finger-versus-heel experiment described earlier. Computer and typewriter keyboards are designed to accommodate the selectivity characteristics of human adult fingers and fingertips. The keys are spaced just wide enough apart so that the width of most (though not all) fingertips fits on just one key and does not spill over onto others, causing the user to create undue errors. When a control site is used that is too large and wide for the spacing or other selectivity characteristics of an array of switches or controls, a situation develops like what you experienced when you tried to type on the keyboard with your heels: You hit other keys, and it slows you down a lot if you are trying to be fully accurate in your keying.

As speeds of activations increase, selectivity also must increase to help ensure that the technology user will hit or tap the intended target and not ones that are close to it in physical proximity. The simplest way to manipulate selectivity for a better match with control sites is to physically spread out the activation sites of switches and/or controls so that there is more space between them on an array. With a simple manual direct selection pointing board using pictures, symbols, or words, this would mean just spreading out the vocabulary items so that there is blank space between them.

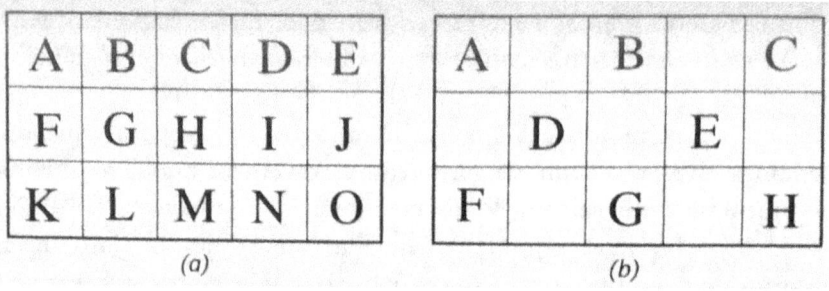

Selectivity of array (b) has been enhanced by creating more space between selection options. As Selectivity has increased, information storage capability of the array has decreased compared to (a).

This same effect can be accomplished on various membrane-adapted keyboards so that a smart phone, tablet or computer can be accessed via a standard QWERTY or modified alphabetical or Dvorak keyboard array. Most VOCAs can be configured with either 2, 4, 8, 16,

32, or another similar multiple of items on the display array, with activation sites growing larger as the number of sites grows smaller. VOCAs can also be programmed so that activation site size stays constant, but neighboring sites are "blank," allowing more spacing and hence greater selectivity for a user who may have difficulty targeting a certain site at desired control site speeds. Selectivity can also be enhanced by electronic means on some AT devices using dwell and averaging as previously described in this chapter; these methods as well as other fuzzy logic capabilities can allow a device to adjust to the imprecise, but repeated selection attempts of a given user.

To mention the rather obvious reverse conditions of the above, smaller control sites in general require less selectivity; switches and controls can be closer together if the body part typically used to operate them is smaller. Consider your car and what it would be like if you had hand controls for the brake and accelerator; They would be awkward and difficult to use if they had the same selectivity characteristics—the same distance between them in this case—as the standard brake and accelerator pedals on the floor that are intended for the larger, wider control sites of feet. Also, from a speed perspective, as the user would move more slowly and carefully to select a given activation site, selectivity, in terms at least of physical proximity, could likewise be compressed. Controls and switches on items where time is not critical in their accurate operation can be quite close together. Check the possible DIP switches on the back of your, router, computer or printer for a perfect object example. Their selectivity for fingers at least is relatively poor, but they can be operated one at a time and with no hurry, so the sacrifice in selectivity for the typical control site employed by most users (fingers, pencil end, etc.) is essentially irrelevant.

In yet another aspect of Fitts' law that we might interpret and apply to our practice in assistive technology, control site mass and muscle force underlying movement can interrelate with sensitivity. In general, the wider and larger the part of the body used as a control site, the less sensitivity in the related switch or control is required for accurate activation. The reverse is also largely true. This makes intuitive as well as practical sense. A switch or control designed to be operated with larger control sites and surrounding larger muscle groups such as foot and leg (i.e., the floor treadle control on a sewing machine, or the swell pedal on an electronic organ) needs to be mechanically "stiffer" and harder to move to and through points of activation than one designed to be operated by a small control site surrounded by a smaller group of muscles (e.g., the keys on a musical instrument that can be operated with all fingers, including the little finger of each hand).

Control or switch sensitivity, selectivity and positioning may match or mismatch muscle force capabilities that technology users can apply with varying control sites.

In assistive technology, the same principles apply. Those who use their neck muscles to move their head from side to side to activate switches will generate more muscle force than someone pushing with a finger or side of a hand. Clients pushing with their knees on a control are using larger muscle groups and a larger control site than if they were to use their tongues as the primary control site, and so on. Though not directly described by Fitts' law, this extension of the concepts embodied in Fitts' central principle seems valid across component areas and various users of AT. It is important for teachers, clinicians, family members, and AT users, if possible, to recognize these potential match-mismatch components related to human factors present in the switches, controls, and interface devices they may wish to use.

Cognitive and Linguistic Effort Factors

These factors in switch and control use also can be viewed from the perspective of Fitts' law and some logical extrapolations from it. Typically, cognitive and linguistic effort tend to increase as target activation sites become more numerous, smaller, and more complex to operate. Each of these parameters can cause switches or controls to become more difficult to operate because of the greater memory, discrimination, localization, and sequencing tasks required of the user. For example, the label array on older but still useful scanning environmental control units that contain several possibilities (10 or so) for control of devices in the home may provide too much information and be too difficult to use for someone who is initially learning to use the unit to only turn the bedroom light on and off.

Similarly, a child or adult who has an expressive symbol vocabulary for AAC use of four items would likely find a full display of symbols across all 128 daunting to use, indeed! In both cases, the number of items for selection should be reduced, with perhaps the size of each enlarged and made as easily discernible as possible. Each user is different, and a one-size-fits-all approach to presentation of switches and controls could be a major clinical error.

Cognitive and linguistic effort factors also may increase for users as individual selection possibilities contained on switch and control arrays become less visually or factually distinct from one another, less selective, more sensitive, or less likely to offer consistent, meaningful sensory feedback to users. In each of these cases, the switches or controls in question may offer confusing, seemingly erratic, or limited communication back to users to let them know what they are (or are not) accomplishing with an AT device. With cognitively young AT users, it would probably be a significant instructional error to present them with a large array of complex, small controls or switches to learn to use. From clinical research, experience, and common judgment, we have come to believe that persons with less well developed cognitive or linguistic abilities need to interact with less complex, less cluttered arrays of activation possibilities.

Children or adults with cognitive, linguistic, or sensory limitations will need less complexity and more redundancy of information to interact successfully with their increasingly more complex assistive technologies across all component areas. Adding tactile markers to a switch or control array for someone with visual or memory difficulties may improve their accuracy of activations. Increasing selectivity so that neighboring items on an array may not interfere as much with a user's memory, sensory, or motor skills may also be of value.

Overall, the reverse of this also tends to be true: Persons who have higher cognitive, sensory, and linguistic abilities can generally operate and interact successfully with control or switch arrays that require more memory, more fine discrimination, and interpretation of symbolic items—and that have less selectivity and more sensitivity. Again, intuitively, more complex AT arrays and interfaces are a better match for users with commensurately more advanced cognitive and linguistic abilities that will guide their motor actions.

Central to cognitive and linguistic organization of switch and control operations in all component areas of AT is the essential usefulness of ordered sets or groupings of information on an array. Quite simply, humans tend to remember things better when they are grouped into recognizable patterns that are familiar to us, as described by Mandler (1984) and others. This makes intuitive, practical sense to us as consumers, teachers and clinicians. From our own learning and teaching, we know that it is easier to find and remember things in an information array if we have some structure, some familiar pattern that we can follow as we view, touch, or hear an array of information. Numerous memory-enhancing techniques rely on this type of technique to help foster memory of names and facts and to enhance retention of information learned in classes, from speed reading, and in clinical uses in rehabilitation and education. Common examples include the "Twinkle, Twinkle, Little Star" melody that helps nearly all children learn and remember the Standard American English alphabet

SWITCHES AND CONTROLS: YOUR PASS KEYS TO TECHN... | 189

at an early age, or the "Thirty days hath September... rhyme that helps us remember the number of days in each month, or even the grouping and patterning of information on an envelope that immediately tells us, by the position and format of written data, for whom the letter is intended and from whom it was sent.

We rely on these types of memory aids to group auditory, visual, and tactile information throughout many tasks in our lives. Their value is perhaps even more essential in AT, wherein we wish to foster the cognitive and linguistic abilities of users who may have challenges and limitations in those areas as they learn to interact with AT devices, and to use the switches, controls, and interfaces that operate their assistive technologies. Elder and Goossens' (1994) have written about the importance of using ordered message sets as opposed to random message sets in arrays of messages used in augmentative and alternative communication.

Which array is more cognitively complex? Why? Which array requires more linguistic effort? Why?

Described a random message set as having no apparent causal or temporal relationship among its member items. An ordered message set, they state, has a logical pattern of semantic interrelationships among items in groups, as well as a logical temporal relationship and sequencing of items relative to one another and to the "unfolding of the natural events of the activity" and AAC use.

Ordered set. *Random set.*

Which picture is easier to remember? Why? Use of ordered sets in displays of information on devices of all types can enhance transparency, natural mappings, and efficiency of use.

Based on work in the field of cognitive psychology and human factors as summarized and described primarily by Elder & Goossens' (1994), and as based on Mandler (1984), Norman (1988), Kantowitz and Sorkin (1983), and Sanders and McCormick (1993), it becomes apparent that for general and assistive technology users, a random organization and presentation of control items can impede the "necessary structure for meaningful recognition, comprehension, and recall" (Elder & Goossens', 1994) of access information. Grouping and ordered presentation in an array of AAC, EC, or other symbols, vocabulary, related controls, switches, and other interface information by affordance, meaning, semantic, temporal, linguistic, cultural, or other constraints and relationships can reduce cognitive and linguistic effort. Grouping and use of ordered sets of information presentation in switch, control, and interface arrays can be significant aids to localization, recognition, sequencing, comprehension, recall, and interpretation of essential operational information.

Elder & Goossens' (1994) describe the example of the bedroom furniture pictures to illustrate the human factors influence of groupings and ordering on recognition and recall of items and information in a message set. As adapted from Mandler (1984), as described in Elder and Goossens' (1994), this human factors principle is illustrated above.

Time

The other component of the Baker equation that we must address regarding Fitts' law is time. Time to activate switches and controls successfully and accurately can obviously be influenced by all the factors we have described here. The potential permutations of interactions and the complexity of this entire process for any given tech user and across users can be enormous and, to a large extent, defy complete empirical measurement across all possible parameters. Much of our work in all these areas, and surely in the aspect of time regarding switch and control activations by our AT users, falls into the clinical judgment and experience areas. Just what works for a given student or client is an incredibly complex mix that cannot always be described clearly or predicted in an objective sense; we must learn from trial and error.

Nonetheless, some generally true statements can be made regarding time. Typically, faster activation speed (how fast the user needs or wants to move a control site into contact with a switch or control to operate it) requires greater resolution of the activation site. In other words, when you need or want to hit a target quickly, and you want to be accurate and successful in activation the first time you try, it helps if the activation site is larger—as large as possible for a given control site depending on the seriousness of need. Faster activation speeds for a given control site also require greater selectivity of the activation site. The switch or control must be distinct, even separated, from similar items that surround it so that chances of the user rapidly activating the site intended—and only that site—are increased.

An excellent common example of this principle is the way powered mobility chairs, carts, and scooters are typically designed. The controls for these devices are usually large, fist-sized joysticks or levers that are mounted in an easy-to-access position for the user. The levers or joystick tops are generally large, with prominent control knobs or flattened pressure spots that afford many users optimal access and control with their hand, fingers, or other control sites. These levers or joysticks also are distinctly separated from any other switches or controls around them so that there is no question from a cognitive, memory, sensory, or linguistic perspective as to what this switch or control is for and where it is on the powered mobility device. It is critically important that the forward-backward, side-to-side control on these devices be operated correctly and promptly, so designs and configurations of these switches and controls have supported the physical, cognitive, and linguistic needs of AT users by making them prominent, distinct, and discrete in operation. The consequences of a user's error activations, such as inadvertently backing over the edge of a stairway when the user wanted to go forward, or misactivating controls while crossing a busy street and swerving or darting into oncoming traffic, could be extremely serious. Making the directional and acceleration controls prominent, distinct, and easy to find and operate correctly, even when a user may be moving a control site in an imprecise manner or in a hurry, is essential in this application.

In AAC, EC, adapted computer access, and play, the safety consequences may be less severe than in powered mobility. Nonetheless, being able to answer a teacher's question quickly in class, or respond promptly to a friends' comment, or play a game competitively with a partner all may hinge on time factors in accessing switches and controls. Variation of resolution and selectivity to meet the individual needs of technology users can contribute to success in all these component parameters of tech access.

SWITCHES AND CONTROLS: YOUR PASS KEYS TO TECHN... | 193

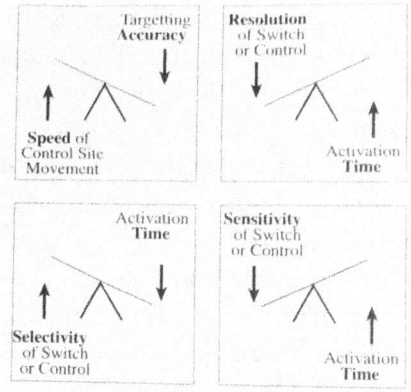

Interpretations of Fitts' law can be expressed as inverse trade-off relationships. Several practical extensions are depicted here. These human factors have numerous important applications across all areas of consumer and assistive technologies.

Closely related to these areas is sensitivity of switches and controls. In many cases, enhanced sensitivity of interfaces may reduce load. Too much sensitivity, however, can foster misactivations and even serious AT operational errors, as could be related to the examples cited here surrounding mobility issues. In some applications, especially in AT used by persons with special needs, a less sensitive switch or control may be the most appropriate, especially one that often may have to be operated by an erratically or spasmodically moving hand, arm, or other control site that may be the best a given user can develop. In ideally designed switches and controls for AT applications, sensitivity will be fully adjustable so that the interface device may be activated by anywhere from minute to extreme amounts of muscle force, as required by the user—and as adapted to different control sites of the user at different times of the day as muscle strength, tone, and accuracy of movement may change. Many AT users who have athetosis, spasticity, or ataxia due to cerebral palsy, or progressive neuromuscular diseases such as muscular dystrophy, multiple sclerosis, amyotrophic lateral sclerosis, have markedly fluctuating motor capabilities from moment to moment and day to day. Matching these types of special needs with switches and controls that have been adjusted appropriately (and are further adjustable) is critical to successful tech use and AT intervention.

On the other hand, when activation movements with control sites can be made more slowly and more precisely, sensitivity can be reduced commensurately. Regarding the time portion of the Baker's basic ergonomic equation, each of the parameters described here (resolution, selectivity, and sensitivity) rests essentially on the opposite end of a teeter-totter from time. In other words, as resolution, selectivity, and/or sensitivity of switches and controls increases, time for accurate intentional activation of switches and controls will tend to decrease. The reverse is also generally true: As resolution, selectivity, and sensitivity decrease, time for accurate intentional activation will tend to increase.

SUMMARY

The essential human factors surrounding successful use of switches and controls are many and complex. With a primary focus on communication skills and elements, this discussion has just begun to address the many detailed aspects that must be considered in the design, manufacture, selection, prescription, configuration, and successful use of these interfaces with consumers, students and clients in the real world. The permutations and possibilities are nearly endless. This area of the field of AT is huge, and several human factors that we take for granted in consumer technologies can be described and understood in ways that follow lawful relationships. These can be clinically applied as well as adapted for special needs users and AT. As always, our focus must be on designing and configuring technology to meet AT users' needs, not on changing the users to fit the mold of the technology we want them to use. This may not always happen—and it may not always be possible for practical reasons—but it serves as a worthy ideal.

This chapter has attempted to present foundational, essential information for a basic introduction to promote understanding and appreciation of the human factors of sensitivity, selectivity, resolution, and several others as they pertain to switches and controls in assistive technology. Our focus has trended toward communication enhancements, yet concepts covered apply across all assistive, enabling technologies and categories of users. Switches, controls, and other related hybrid interface devices are indeed the pass keys to the rest of assistive technology and to the world of inclusion for special-needs users. If these pass keys are improperly designed and configured, they can present major barriers for general users of technology, as well as deserving potential AT users. If they are properly, appropriately designed and configured, they can indeed become like Archimedes' lever, allowing even users who have limitations or special needs to move their worlds of play, learning, mobility, work, and inclusion in life, becoming more independent through access to all component areas of assistive technology.

REVIEW QUESTIONS

1. Define switches and controls. Describe several of each in your daily life, and several that most technology and AT users may encounter each day. What are their specific functions? How are these alike and different across types and applications?
2. Cite examples from your own life and use of devices for each human factor mentioned so far regarding switches and controls. What types of switches and controls are rare for you?
3. Which human factors have you found helpful and useful in the switch and control technologies you use? Which human factors have not been part of the technology you use?
4. Regarding your own students, clients, or patients, which human factors pertaining to switches and controls have affected their use of technologies, particularly AT? Why?
5. Cite specific examples of changes or improvements to better incorporate human factors we have mentioned that you would suggest with tools/devices you have used or currently use with consumers, students, clients, patients, or family members.
6. What adaptations did you or the client have to make because of human factors limitations you found in these devices?
7. Cite three examples each of Fitts' law applications in your own daily use of technology across the areas of physical

effort, cognitive/linguistic effort, and time load. Discuss them regarding your car, your home appliances, and your recreational devices or technologies.

8. Cite specific examples of how Fitts' law principles and applications may impact students, clients, or patients in daily use of simple and complex technologies.
9. What is the most opaque, poorly designed technology or device with which you or your clients have ever attempted to interact? Specifically why? Which human factors "violations" can you identify in these items?
10. What is the best, most transparent technology that you yourself or your clients have ever attempted to use? Specifically why, from a human factors perspective? What was it about the switches and controls that made it so?

7

SCREENS: YOUR GATEWAYS TO TECHNOLOGY

Technology, like art, is a soaring exercise of the human imagination.
Daniel Bell

Innovation is the outcome of a habit, not a random act.
Sukant Ratnakar

Whether referred to as monitors, displays, arrays, or screens, these visual gateways provide entry into successful use of many consumer and assistive technologies. Each allows us to see what we are doing, and to know how we are interacting and functioning with the devices we and our consumers, students, clients, or patients require. For our purposes here, we will simplify, using the term *screens* to refer to all. Visual portrayals via screens of alphanumeric characters, icons, pictures, plus other graphics and data via any of these portals allow us to fully interact with the assistive and other devices we rely on for a variety of purposes.

The function of screens is so basic to many complex technologies that we tend to take it for granted. Imagine how difficult it would be to type an email or play a video game if you could not visually monitor results of your keystrokes on a keyboard, or the effects of the movements of a joystick or other switches or controls. Screens are essential to our successful use of many general consumer technologies; they are also essential to our clients' and consumers' use of AT, even for many clients who live with visual limitations. Nonetheless, the dominance of graphical user interfaces, GUIs, convenient for most sighted users, has indeed presented obstacles for persons living with vision impairment or who are blind. Various workarounds such as speech and auditory readback, embossed or raised printing, vibrotactile Morse code, refreshable Braille pin readouts, and other alternative signaling methods do help, but still cannot offer the consummate ease and efficiency experienced by sighted users with high-quality screens.

In assistive technology, screens are found as essential components of phones, tablets, entertainment tech, AAC devices and adapted-access computer systems. They may also be included in some adapted toys, mobility devices, and ECUs, and likewise be included in devices combining several AT component areas and functions. Sophisticated combinations of AAC, EC, and adapted computer access systems with screens are common. Screens are also found in smaller forms on many powered wheelchair configuration controls and with stand-alone environmental control units, although these screens may be simpler alphanumeric displays of information or arrays of indicator lights.

Screens currently included in many popular consumer and assistive technologies are of these main types:

Liquid-Crystal Display (LCD): These flat, backlit screens are part of many smart phones and tablets, as well notebook, laptop and desktop computers, plus toys, calculators and other devices.

Light-Emitting Diode (LED), organic light-emitting diode (OLED and active-matrix organic light-emitting diode AMOLED): These flat, self-lit screens are found with many devices, and are popular for applications in consumer and assistive digital technologies. They are often used for alphanumeric displays on control panels of appliances, clocks and other data readouts on various common devices.

Plasma: These large, flat, brilliant screens offer excellent picture quality for motion, including sports, movies and video gaming, but plasma screens are less used now because of cost, weight, screen burn-in, and limited lifespan.

Cathode Ray Tube (CRT): Once the primary monitor for home television sets and computers, CRTs are still used in some specialty applications, but have largely disappeared from the consumer and assistive technology markets.

LIQUID CRYSTAL DISPLAY (LCD) screens are much thinner and lighter in weight than CRT screens. They are now found as integral components of many compact, portable electronic equipment items that have a visual screen display of complex graphics, text, and other information. LCD screens allow the monitor portion of a device to be essentially flat, with only a few millimeters to about one centimeter of actual screen depth or thickness required, including the protective casing material. In the past, LCD screens have been available only in variable gray scales, such as those found on pocket calculators and other such devices. These LCD screens are not backlit but, rather, have a mirrored surface behind the LCD display to reflect ambient light (DiChristina, 1997).

More recently, however, active and passive color matrix LCD screens are considerably improved, and they offer a variety and brilliance of colors that rival those of older CRT displays. For examples, Apple iPads offer excellent multicolor, multiuse LCD screens, whereas Kindle devices use E-Ink, an LCD screen superior for reading just text, and particularly useful in bright lighting such as outside. Current versions of many portable devices incorporate LCD screens for as they are way lighter and thinner, and require less voltage and current to operate, than CRT screens. LCD screens are generally suitable, adaptable, and durable for potentially robust portable applications. This is especially true where mechanical vibration, jostling, or bumps may often occur—exactly as is encountered moment by moment with many forms of assistive technology used in real life by consumers, students and clients at home, school, and work.

LCD screens function via a unique and complex process. Liquid crystal is a material that, within certain temperature ranges, may flow like a liquid while also having certain properties that resemble those of a physically solid crystal (Bernstein, 1989; DiChristina, 1997). LCD screens work by drawing electrons across a flexible palate or substrate of liquid crystal material spread over the top of transparent electrodes that power the device. Liquid crystal is a highly formable compound made of small chemical crystals that can be applied in thin coats to transparent materials, such as the powering electrode surface of the LCD screen. When electrons from the underlying power panel are caused to flow through the liquid crystal material, the molecular bonding patterns of the LCD layer can be made to change their orientation (DiChristina, 1997). Their altered path allows light to pass through variably (or not) to define an image.

Overlying the liquid crystal substrate in the screen "sandwich" are separate regions of alignment layers, RGB (red, green, and blue) color filters, glass electron filters, and polarizing light filters, which together operate to control the color, focus, and brilliance of the screen display. These layers above the liquid crystal substrate control and modify the light that is permitted to pass through them by the constantly realigning liquid crystal molecules that lie immediately below (Bernstein, 1989; DiChristina, 1997). Monochrome LCD screens work in a similar manner but do not have the RGB filters. To add brilliance, both color and monochrome LCD screens on VOCA devices, notebook computers, and other devices can be actively backlit from a separate light source below the screen so that screen contrast and luminance in higher ambient light surroundings are enhanced.

A variation of the LCD screen, the touchscreen, is commonly found as part of computer screen displays, personal data devices, and as add-on peripherals. LCD-based touch screens have internal layers that are coated with a transparently thin metallic film. Alternatively, they may be embedded with a network grid of extremely fine, essentially invisible wires. When the screen is touched and depressed by a finger, screen pen, or other selection device, the low-level voltage flowing through the film or wires is altered. The amount of voltage change is determined by the proximity of the control fingertip and the distance of the touch from the screen edges, with greater voltage change occurring as the screen is touched farther from its borders. The microprocessor interprets the voltage changes from touches in different screen locations and translates the information into X and Y screen coordinates. These voltage changes and instant mappings of coordinates allows the computer to recognize where the touch has occurred, to identify what it signifies in terms of screen data displayed below the touch, and to track the direction of movement of the pen or finger by constantly replotting X/Y coordinates (adapted from Wessner, 1997).

In general, LCD screens operate with low current drain (a few milliamps), and produce virtually no heat in their operation—quite different from CRT screens. Plus, the truest color LCD screens are of the active color matrix type and are capable of near-CRT quality in brilliance, focus, and overall sharpness of image (DiChristina, 1997; Grabowski, 1997). The greater brilliance of active-matrix color LCD screens also makes them more resistant to glare from external light sources. All small portable computers, calculators, and other screen-display devices that must portray complex, intricate images or data increasingly use LCD and touchscreen technology. The compactness and ruggedness of LCD screens of all sizes and types, and their increasing clarity, brilliance, and color definition make them suitable for many applications across all component areas of assistive technology. Their overall portability and durability make them well suited to be part of the everyday tools of people's lives.

A quick inventory of home, school, recreational, and work settings will reveal how common and important screens of all types are to all of us. In our homes, screens of all types are found on televisions, computers, phones, tablets, calculators, entertainment devices, smart watches, cooking stoves, microwave ovens, and other consumer, assistive and medical devices. Several also are probably integrated with other items that we have overlooked because they are so much a part of our lives. Think of all the screens that you use in your home, school, work, and other settings in your own life; you will likely be able to list many. In our cars, now, it is standard to have screens that portray and display data on the front dash or control panel and in other locations. These screens convey data from speed, fuel, and oil levels and engine temperature to phone calls, frequency of radio stations, music selections, time, and passenger compartment or outside temperatures, among many other environmental readbacks. Screens are everywhere. We cannot live without them in this industrialized society, and neither can users of AT.

In a few pages, we will look further at some essential human factors we must consider that are related to screen design, function, use, and selection for our students, clients, and patients who rely on these gateways to assistive technology. The writing of this book, and particularly this chapter, was motivated in large part by the many experiences we have personally had or known about from colleagues regarding children and adults in AAC and AT diagnostic sessions. Client and screen characteristics can interact in ways that we often do not consider and can impact our diagnostic and intervention attempts with assistive technology.

Every so often a student or client will perform poorer than expected on an evaluative task that involves an AT device with a screen display. Unfortunately, this poor performance is commonly attributed to the potential AT user, and to his or her own personal cognitive, sensory, or physical limitations and disabilities. Although these client characteristics may indeed influence performance in many cases, in our own experience and that of other colleagues, other human factors variables may be responsible. Sometimes seemingly simple elements of the AT diagnostic session, such as the screen size or resolution, or the glare or distracting reflectivity from a screen display of an AT device, may alter a person's ability to respond to and interact successfully with that device.

Glare, reflectivity, and other human factors related to screen displays across all relevant AT devices can become, despite their apparent simplicity, significant confounding elements that can inaccurately skew assessment, evaluation, and diagnostic results. This may be particularly true in our work with consumers and clients with severe sensory, motoric, and communication limitations who may not be able to notice, report, communicate about, or adjust to these factors on their own. It may become even more critical if we, as sales and support personnel, family, and professionals, are not aware of and controlling for these factors in our awareness, diagnostic or instructional sessions. These and other essential human factors surrounding screen displays may go unaddressed completely, and may have serious effects on use of consumer as well as assistive technologies by individuals. More details will follow, after this discussion of other screen types.

LIGHT-EMITTING DIODE (LED, OLED and AMOLED) screens have become practical, useful and popular types of displays. LED screens show letters and number characters in brightly illuminated red or blue colors. LED screens, once used mostly to display only alphanumeric information (letters and numbers) and now used widely used in complex imaging applications in consumer and assistive technology today, largely exceeding uses of CRT and LCD screens. They are so numerous and common in our home, school, work, and recreational environments that often they blend into the background of these settings. We take them for granted— yet they offer the primary means by which we interact visually with a huge range of important consumer and assistive technologies. In assistive technology, LED screen displays are also found with the control configuration units of powered wheelchairs, some EC control modules, and with clocks, thermometers, and some wristwatches. Once limited to simpler applications, LED, OLED and AMOLED screens are now suitable displays for many consumer and assistive. They have become part of our daily lives in technology-heavy societies.

---

100 words about *LED, OLED, AMOLED AND FLEXIBLE SCREENS*

Light-Emitting Diode (LED) screens are often used outdoors because of their brightness, relying on individual light-emitting diodes as pixels to compose images. First produced by Samsung, Organic Light-Emitting Diode (OLED) displays are LEDs enhanced by addition of a thin organic compound film that illuminates when activated. Not requiring a backlight like most LCD screens, LED and OLED displays produce their own light when a current is passed through them. Improved AMOLED screens (Active-Matrix Light-Emitting Diodes) create better-quality images and allow

thinner phones drawing less power. **AMOLED** displays are becoming less expensive and more widespread. Flexible screens that can be bent or rolled without distorting images use OLED, LCD or Gyricon tech in consumer and assistive devices. See https://www.trustedreviews.com/news/what-is-oled-3285263 and https://en.wikipedia.org/wiki/AMOLED

PLASMA screens, large, bright and heavy, were once popular. Regarded for a while as replacements for CRT and LED screens, they seemed omnipresent in restaurants, bars, schools, and family TV rooms, and were enjoyed for viewing sports, movies, performances and concerts. They also worked well for playing video games, especially with groups of players. Supplanted by video projection systems and improved, expanded LED capabilities and options, plasma screens have declined in acceptance and use. See the inset below for a condensed summary and selected online references.

---

100 words about *PLASMA SCREENS*

Plasma technology allowed large flat-panel screens, often in the 40-to-65-inch size range, that could be mounted on a wall. To produce a picture, a mixture of mercury with noble gases was contained between two sheets of glass, then activated with an electrical current, turning the gasses into plasma. The result was a clear display, well suited for fast-action events such as gaming and sports. Offering excellent picture quality, plasma screens featured reduced blur of motion, viewing angles wider than many other screens, good contrast ratios, and supported higher resolutions of 1920 X 1080 pixels. However, with their heavy weight, loss of brightness over time, and screen burn-in yielding a lifespan of about 10 years, plasma displays have fallen out of favor. Panasonic discontinued making them years ago. Samsung made a similar announcement in 2014. Expensive, heavy, and costly to operate, plasma screens are now largely passe. See https://www.computerhope.com/jargon/p/plasma-screen.htm and https://en.wikipedia.org/wiki/Plasma_display

---

CATHODE RAY TUBE (CRT) screens were, for decades, the most frequently encountered displays for most mounted or placed applications such as home televisions, desktop computer workstations, video monitors for education and entertainment, and commercial displays of constantly updated information in businesses. They are still in use in some sectors of the market and some places in the world. CRT screen monitors are relatively large, bulky, and deep (front to back dimension). They operate by means of one or more electron "guns" at the rear of the picture tube rapidly scanning beams of electrons across the back side of the curved front glass screen coated with phosphor dots. Each phosphor dot glows for a short time after a "hit" from the electron beam, with the color of the glow being controlled by the type of phosphor dot on the picture tube surface (red, green, or blue glow in a color monitor; gray scales in a black and white or monochrome monitor).

CRT screens are usually found at the front of a relatively deep, bulky monitor case because the electron gun must be mounted at a distance from the screen toward which it will direct its scanning beams. It is difficult to make CRT screen devices compact because of this depth required for operation of the scanning beam. Operation of a CRT screen also consumes substantial electrical current, much of which results in production of considerable heat. Some authorities have reported that a functioning CRT screen in a room produces about as much warming effect on the room as a human body (Grant & Brophy, 1994).

100 words about *CATHODE RAY TUBES: ARE THEY STILL IN USE?*

Yes. Cathode Ray Tubes are still being made and used. They have retained usefulness in special airline and vintage gaming applications, among others. Boeing, Airbus and Lufthansa relied on CRT monitors which can perform with better lag than OLED and LCD screens. In some parts of the world, CRTs are still preferred because they are in already in place, and will continue functioning without having to upgrade. Durable CRTs can offer clearer images at several resolutions for popular video game competitions without distracting center screen blurring. See www.thomaselectronics.com/faq https://en.wikipedia.org/wiki/Cathode-ray_tube

Despite some of their disadvantages and decline in use, CRT screens have proved over the years to be reliable and durable. They are generally easy to view under almost all lighting conditions because of the intensity of the glow they emit (Ostrom, 1993; Grant & Brophy, 1994). Most CRT screens, whether their display is in color or monochrome (black and white or gray scales, black and green, or black and amber), are also readily adjustable for brightness and contrast to match room conditions and the needs of the users. Cunningham (1980), Ennes (1971), and Sherr (1979), among others, have written extensively about CRT screen displays, and should be consulted for more detailed technical information.

SCREENS: YOUR GATEWAYS TO TECHNOLOGY | 207

CRT screens are still in use. LCD and LED screens are found with a greater variety and scope of consumer and assistive devices.

---

100 words about *BACKGROUND AND HISTORY OF THE CATHODE RAY TUBE (CRT)*

Cathode Ray Tubes, now topics for boat-anchor humor and bane of landfills, charity stores, and recycling centers, have been reliable gateways to commerce, entertainment, socialization, and education for decades, with roots into the 1890s. From earlier shared research, Braun developed the initial CRT in 1897. He was first to envision it as a display device. "Distance electric vison" of the Braun Tube was described in the journal Nature in 1908 as a means to send and receive information. Continuing research into the 1920s explored CRT use. Pioneering efforts by Takavanagi,

Farnsworth, Zworykin and others led to early TV by 1935. For additional detailed developmental history and technical explanations, please see: https://en.wikipedia.org/wiki/Cathode-ray_tube

---

HUMAN FACTORS RELATED TO SCREENS

The number of human factors variables to consider with screens is surprisingly large for such an apparently simple topic. Although we take screens for granted in many aspects of our daily lives and in AT use, the details of our interactions with them are highly complex. We will attempt to address some aspects of this interaction that seem most relevant to AT applications, while readily acknowledging that additional, more complex issues and data exist in this area. These can and should be addressed from a more sophisticated ergonomics and human factors engineering focus, as has been offered by Anderson, M. (2022), Grant and Brophy (1994), Ostrom (1993), Sanders and McCormick (1993), and others. As mentioned previously, a listing of excellent supporting references is included in the bibliography. Our efforts here, as is true throughout this book, will be to synthesize and focus on practical, clinically relevant insights and information derived from a variety of sources and experiences. This information can then be readily used by consumers, teachers, therapists, families, and others who must help select, acquire, and maximize use of AT for their child and adult clients who live with special needs.

---

100 words about *SOUND, HEARING, AND ENHANCED AUDIO FEATURES*

Screens often involve sound. And hearing loss can affect users of all ages. Presbycusis, the sensory neural hearing loss of aging, and other conductive or sensory hearing impairments, are common, reducing perceived intensity & clarity of speech, voice, music & other sounds. For partially sighted & blind users, audio

> depiction of on-screen events is a primary way to interact with programs, games, movies & media. Although corrective optical lenses can intensify & clarify visual input, hearing aids & amplification, useful for many, mostly make sound louder, not clearer. For users with various hearing, vision or other sensory challenges, enhanced audio depiction/description, plus visual captioning for some, can help clarify dialog, distance, position, action, plot & meaning. See https://verbit.ai/assistive-technology-for-visually-impaired **and** https://www.nidcd.nih.gov/health/captions-deaf-and-hard-hearing-viewers

---

In addition to sound, hearing, captioning and enhanced audio features, essential human factors related to clinical use of screen displays may be grouped into these three categories:

- **Human factors related to the physical size and bulk of the screen**
- **Human factors related to the display quality of visual information on the screen**
- **Human factors related to user proxemics, seating, and positioning for the screen.**

We will address each of these in turn and will attempt to summarize basics of this this large, important topic within this chapter. There is far more to say than what we will cover here. For more information, the reader is referred to the selected references section of this book.

Physical Size and Bulk of the Screen

Human factors that relate to the physical size and bulk of the screen are particularly important in AT applications. Not only are screen display devices in AT commonly used by persons with special physical needs, who may have reduced strength and motoric capacities to lug or move equipment, but also these screen-bearing devices are often subjected to more jostling, pounding, and jarring than general consumer technologies because of the motor inefficiencies common to many of their users. Most CRT screen devices in AT applications are not found in portable devices; they may still be part of desk- or table-mounted applications and stationary video equipment because of their size and bulk: A typical CRT computer monitor weighs approximately 20 to 25 pounds, making it much too heavy to move often, or to carry around as a portable display device. Despite their size and bulk, and their seeming durability for many general consumer applications, the types of screens used in AT applications can be subject to frequent, significant jarring and shaking—much more so than in most general consumer uses. Such physical trauma to the CRT screen device may be caused by users with reduced motor control and precision as their wheelchairs, other mobility aids, or their own limbs or bodies bump and shake the computer CRT monitor or supporting monitor stand. Some AT users with pronounced extensor reflexes or other powerful, quick involuntary or imprecise movements may significantly jostle the monitor and screen during use. Despite their general ruggedness and ability to absorb abuse, repeated jarring of CRT screens can indeed break or disrupt delicate electronic components and solder joints. Repeated ballistic vibration of the CRT screen can cause misalignment of the electron beam that must scan precisely across the back of the screen many times per second to make the screen glow with accurate, focused images.

Taking reasonable precautions to shield a computer workstation, mounting systems or video display terminal from frequent, excessive bumping or jarring should be a commonsense preventive practice. It can assure that the screen-based device being used does not receive an overabundance of sudden, quick vibrations. Using tables and stands that can be individually adjusted to a given user's needs for access in sitting, standing, or lying positions can be a major factor in preventing undue mechanical trauma to equipment (and user). Padding of desk or table edges and legs can also be of help. Often, the OT, PT, family, or others can provide instruction and can practice with AT users on smoother, less disruptive entry to the workstation. They may also help users learn to restrict, restrain, or redirect in more useful ways certain strong reflex movements that might affect the screen and other equipment they are using.

Consultation with OT, PT, and technology professionals may reveal practical, transferable ways for clients to move up to any desktop screen device in a gentler or more precise manner (rather than ramming forcefully into it each time, as we have seen with some clients). OT and PT professionals also can help AT users practice utilizing gentler, less ballistic movement patterns of limbs, torso, and head for CRT device access and control. They may help AT users to pattern ways of managing reflex movements during times of excitement or surprise that may cause less jarring vibration to the CRT device, whether it be a desktop computer, television/DVR combination, video game terminal, or other sensitive electronic devices.

Physical vibration and jarring are less pertinent with LCD and LED screen devices than with CRT screens. This, plus the fact that LCD screens are vastly lighter in weight and more compact in physical size, make LCD screens the clearly preferable choice for portable applications in assistive technology. Like the many consumer electronic devices mentioned earlier in this chapter (such as portable television sets and computers), the screens on portable VOCAs, ECUs, mobility device displays, and adapted toys used in AT nearly always involve LCD screen components. This is so for several reasons related to weight and size of the screen, durability, adjustability, and maneuverability.

The weight of a typical LCD screen from a VOCA or notebook computer is often in the 0.5- to 0.75-kilogram range (or less), lending itself much more readily to being carried by an AT user. This is especially true for someone who is physically less strong, less motorically coordinated, or smaller in stature than others. The relatively light weight of LCD screens also allows such devices to be mounted on wheelchairs and other mobility devices with ease and safety. The control electronics portion of the computer or VOCA, not the LCD screen, contributes the primary weight and bulk in this case.

Additionally, most LCD and LED screens are remarkable in the amount of physical jarring and abuse they can withstand. Because they are composed of softer, more flexible materials, LCD and LED screens do not shatter or break as a CRT screen would. Being dropped on the floor, exposed to extreme heat or cold, or struck by a hand, foot, toy, or other object typically causes less harm to these screens. Typically is the key word here. These devices are, of course, not indestructible under all possible conditions.

For example, as demonstrated by Deming (1996), Friel (1997), and others, the LCD screen of some portable VOCA devices can withstand strong, repeated pounding by an adult male or others. Even when in full operation and when configured for ongoing communication, the LCD screen of these devices can be struck remarkably hard and repeatedly with no damage to the unit. LCD screens also are sealed so they can better withstand humidity, moisture from drooling and mealtime use, plus overt spills of liquids. This type of ruggedness is essential when a device is to be used by children or adults as they integrate into the real world of activity, mobility—and of taking their AAC and other AT devices with them wherever they go. Gravity, falls, mishaps, and other kids tend not to respect how nifty AT is. The technology must be built to withstand potential rigors of use and misuse in daily life. LCD and LED screens can contribute markedly to this need for ruggedness.

Despite all these advantages, however, early LCD screens do indeed present some significant disadvantages. These included reduced overall brightness and reduced control of brightness and luminance ratios when compared to CRT screens in high ambient light environments. This is not always easy to detect or account for when such devices are being used with children or adults who may have reduced abilities to sense and/or convey that screen dimness is affecting their use of a device. LCD screens also demonstrated a definite propensity for reflective glare from ambient light sources, a significant contributing factor also not always readily detected or communicated by AT users who have special sensory or communication needs. Many improvements have occurred. LCD and LED screens are standards of excellence.

In the case of CRT screens, most that were integrated with desktop or tabletop devices were positioned in an upright manner with the screen parallel to the overhead lighting sources commonly found in schools, clinics, and hospitals. In these cases, the CRT screen face was perpendicular to the user's line of sight from a sitting or standing position, with the viewing angle generally fixed. That is, the weight and bulk of CRT screens required rigid, firm, fixed positioning and mounting so that the screen display was safe and could not break in a fall—or fall onto someone. Although some amount of positional adjustment was possible with many off-the-shelf CRT computer monitors, the range of vertical motion was generally limited to a few degrees of tilt up/down or forward/backward. This may not have been enough to change glare-related angles of the screen surface adequately. Conversely, LCD screens, because of their light weight and trim physical construction, continue to afford many options in mounting, positioning, and adjustment. What they once lacked in brightness and clarity compared to CRT screens have been resolved, and is further enhanced with their ability to be easily repositioned to compensate for glare and reflectivity of ambient light. These and other human factors related to viewing of CRT, LCD and LED screens are further addressed later in this chapter.

Screen Quality of Visual Display Information

Human factors related to the display quality of visual information on the screen will be considered next. This is the area of human factors related to screens that perhaps receives the greatest amount of attention, research, product development, and clinical effort on the part of teachers, clinicians, and family members. Many simple and complex variables can be addressed in this topic area. We will cover some of the more clinically relevant items. These human factors-related variables include but surely are not limited to the following areas of consideration across all AT devices that incorporate use of CRT, LCD, or LED screens:

- **Contrast and brightness of screen, resolution, flicker, and focus**
- **Screen glare and reflectivity**
- **Font sizes, types, contrast, and labeling**
- **Screen feedback for entry errors, misspellings, and other device functions**
- **Static and dynamic displays of text, graphics, symbols, and icons**

Contrast, Brightness, Resolution, Flicker and Focus

We will first consider several aspects of screen contrast. Most basically, a CRT, LCD, or LED screen display must appear bright and distinct enough from its surrounding case, equipment, and ambient light environment (whether indoors or outdoors) to be discernible to the user. In short, a user must be able to locate the screen before they can use it. This basic, central concept is known as the luminance ratio of the screen to the ambient light environment. It may seem obvious or trivial to most of us with full vision. Yet it becomes particularly important with children and adult users who have special needs, particularly extremely low vision. Visual acuity of AT users or candidates, within the diagnostic range of only sensing light perception and the direction from which light comes, known as projection (Lorenz, 1997), would necessitate that her or his supporting personnel pay attention to the orienting of the user to the presence of the screen display as a first step toward eventual AT use and interaction with a screen display device.

Although seemingly basic and evident, this first step must not be ignored, particularly with cognitively young, low-vision students or clients who may also have physical limitations that could interfere with or prevent their ability to adjust the screen brightness, surrounding lighting, or even their own posture and orientation to the VOCA, computer screen, video monitor, or other display devices. Their eventual ability to discern information from and interact with a screen must derive from their initial ability to locate it. Sometimes, in extreme cases like these, the addition of tactile and/or auditory markers to aid in their orienting to screen devices can be of value for some whose ability to discern screen contrast, projection, and brightness is limited.

This primary aspect of contrast, luminance difference or luminance ratio, is described well by Grant and Brophy (1994) and Sanders and McCormick (1993) as the comparison in brightness between the glow of the screen and the surrounding ambient light environment. The surrounding light environment could include the reflected brightness of nearby walls, other equipment in proximity, and any paper document that users may also be viewing or working from while using a screen display device (typing from printed copy, etc.).

Luminance in our use here refers to the brightness of the screen itself and is the primary element in the contrast of the screen with ambient lighting and reflection. Grant and Brophy (1994) further state that achieving a luminance ratio of background lighting to screen luminance for persons with normal vision of no more that 10 to 1 is desirable (i.e., the background, surrounding light environment should be no greater than 10 times as bright as the screen luminance when empirically measured). More exact luminance ratios for persons who have a variety of special needs, as is common with many AT users, must be defined.

Grandjean (1987) (cited in Grant and Brophy, 1994), recommends a conservative 3-to-1 ratio, a less drastic comparative difference between background lighting and screen brilliance. Overall, Grant and Brophy suggest that a comfortable luminance ratio can be achieved by keeping ambient illumination relatively low, at around 200-500 lux (20-50 footcandles). This may be particularly true with dark background screens. They further state that screen brightness should be adjusted according to ambient luminance of the user's room or workspace, and that use of light-background screens makes it easier to achieve usable, optimal luminance ratios.

Although the need for this type of adjustment may be easily sensed and accomplished by most persons with normal-range abilities, those AT users living with severe limitations and disabilities, as mentioned above, may find this a daunting or impossible task. Teachers, clinicians, and family members need to be alert to practical aspects of luminance ratios of screens, and their impact on overall screen contrast. They should be alert and ready to adjust pertinent variables with a user's AT screen display device, or lighting within the AT user's home, school, or work environment as needed to achieve optimal luminance ratios and resulting contrast of AT device screens with background, ambient luminance in different indoor and outdoor settings.

Other essential considerations, especially regarding screen selections for use in AT, are resolution, flicker, and focus. In our use here, the term resolution refers to how sharp and clear the image is on screen, that is, how readable the information displayed on the screen is to most viewers. Screen resolution is directly related to the density of pixels per square inch or square centimeter of the screen surface. A pixel, a compression of the phrase "picture element," is the smallest component area of a display that is individually addressable. Pixels are the individual cells of phosphor dot material on the back of a CRT screen that can be made to glow by the scanning electron beams, or by other lighting means with LCD and LED screens. Resolution can be expressed as the number of pixels that occur per scan line horizontally as compared to or multiplied by the number of vertical lines on the screen per square unit measure (i.e., 640 by 480 or 1,024 by 768, and so on), per Grant and Brophy (1994). The more pixels per square measure, the higher the screen resolution; the higher the screen resolution, the clearer and sharper the pictures, images and other information-bearing characters will appear on the screen.

Grant and Brophy (1994) and Ostrom (1993), plus our own experience and that of many others, suggest that although we typically desire higher resolution monitors for our own use and for use by students and clients, the higher resolution capabilities of these screens may produce smaller (though very clear) images. Ironically, because these clear, sharp images are made smaller on screen, they can become harder to see and to read for persons with low vision than if they were reproduced on a lower resolution screen. Certain graphics capabilities with newer software and screen types also may produce this effect. Despite our best efforts to create highly visible, readable data and graphic displays for our clients, this effect can render a screen display unusable for some AT users. We have found that adjusting the higher resolution monitor to a lower, less sharp setting often can allow software of this type to be used. This adjustment produces on-screen images that appear "fuzzier," but the overall increase in size of the image on screen may allow for greater readability by the AT user.

Flicker is an important element of CRT screen function related to its refresh rate. This refers to how often the scanning electron beams sweep across the back of the CRT screen to produce a glow in the phosphor dots. Screen brilliance briefly dampens and then brightens with each refresh cycle, known as the critical flicker frequency (CFF). CFF for general-application CRT television monitors is typically in the range of 30 to 50 Hertz (cycles per second) depending on the make of monitor (Sherr, 1979). CFF for computer CRT monitors was about twice this rate.

When viewing a CRT screen from directly in front, most people cannot detect the flicker of the image as it is refreshed or updated by the scanning beams 30 to 50 or more times each second. The rapid fluctuations in screen brightness blend smoothly together. This is due to the greater concentration of cone cells near the centers of our retinas. They are less sensitive to light variations and contrasts than the rod cells, which are located throughout our retinas, including along the sides. However, when we view a CRT screen on an angle, from the sides of our eyes, nearly everyone can pick up the flickering in screen intensity. The differences in the light detection sensitivity of our rod and cone cells become readily apparent to most viewers, making screen flicker distracting, annoying, and even physically unsettling for some persons during even brief side viewing. LCD and LED screens of all types do not demonstrate this problem.

Consider the impact of this greater perception of screen flicker on someone who is positioned or seated in such a way that he or she must always view the screen from a side angle; this manner of viewing can be markedly counterproductive to successful use of the screen device over extended time periods for work or recreation. Flicker also can be distinctly dangerous: For some seizure-prone individuals, regular, pulsed fluctuations of screen brightness (or other lighting) can induce seizures (e.g., Lorenz, 1997; Sanders & McCormick, 1993).

The rate and intensity of flicker for CRT screens, if still used, must be designed and adjusted to be minimally distracting for all users, including those with special positioning and viewing-angle needs. CRT screens should be viewed by AT users in a direct front-on manner whenever possible to reduce the likelihood of flicker becoming a significant distraction or even a hazard to the user's health. As the flicker of screens, like that of other luminance sources, can trigger seizures in some viewers, teachers, clinicians, and family members, as well as AT users themselves, must be especially cautious about flicker when an individual is seizure prone. The monitor or other flashing light source should be repaired or replaced immediately, and the user should stop viewing the screen as soon as inordinate flicker is noticed.

Several authors have pointed out that users' perception of screen flicker can also be exacerbated by its interaction with ambient fluorescent lighting, lower quality monitors that have a slower refresh rate, and feedback signals such as flashes, blips or other screen changes and operational "signals" induced by software controlling the CRT screen computer monitors (Grant & Brophy, 1994; Ostrom, 1993; Sherr, 1979). These must be noticed and controlled for with users who have special-needs, and who may not be able to indicate discomfort in their viewing of CRT screens. Because of the method of function of LCD and LED screens, flicker is not a relevant human factor for them. These screens operate with a constant general glow that is not activated by a scanning electron beam, as the entire screen lights up as voltage is variably applied.

Focus of a CRT screen refers to how blurred characters or images of any size appear. While some of this is related to resolution characteristics of the screen (density of pixels per square unit measure) as previously mentioned, focus also can be affected by internal problems in the monitor or by the adjustment of settings both outside of the monitor on the control panel of the case, or inside the monitor. External adjustments for focus can and should, of course, be modified by users or their support persons to attain optimal balance of contrast and brightness for the best possible focus. Commonly, just turning a few knobs or pushing a few buttons on the control array of the monitor can help make significant changes in focus that will suit the needs of users. Internal focus adjustments to the CRT monitor can also be made but should definitely be left to a skilled technician. Correct tools, training, and skills to make these sensitive adjustments, particularly when working around the potentially lethal voltage and current levels found in most CRT screen monitors, the province of electronics professionals only.

Other factors beyond the electronics of the screen itself can affect focus. Dust, dirt, and saliva or food smears on a screen may significantly affect the clarity of screen images and should be wiped from the screen as often as needed. This is particularly true with AAC users who always have their screen display device with them, typically in a dirt-receptive horizontal position. As these children or adults may eat, drink, drool, spill paste or paint, or otherwise clutter the surface of their screens, they or their support persons should be alert to the need for regular cleaning of the screen. Those who use AAC should learn to communicate the message "Could you clean my screen?" as a useful interactive safeguard so that they can request cleaning as needed, especially if they cannot do this by themselves.

Focus can also be a function of the AT user's visual capabilities. If a CRT or LCD screen cannot be adjusted to meet users' visual focus needs, consultation with an eye care professional is wise. Ophthalmologists and optometrists can conduct valid, meaningful evaluations of visual abilities even with patients who are severely limited in communication, cognition, or mobility (Lorenz, 1997; King, V. H., 1997). These eye care professionals perform a passive retinoscopy, which is an objective means of determining the patient's refractive error or need for corrective prescription lenses.

Retinoscopy can be performed with the patient sitting, lying, or standing (although positioning of the patient for this is not always easily accommodated by all practitioners). It can be completed by the eye care professional through the viewing and measuring of dimensions and characteristics of the patient's eye structures. Interactive abilities are not necessarily required on the patient's part for him or her to produce meaningful results for possible vision improvement via the most common of assistive technologies—corrective lenses. We have observed over the years that consumers, families, teachers, or clinicians often believe that their student or client is too disabled, too limited in abilities to participate in a reliable, meaningful vision examination. In the care of skilled vision professionals, this does not have to be the case, according to Lorenz (1997) and King, V. H. (1997). Improving individuals' abilities to focus visual information from a CRT, LCD or LED screen, as well as their overall visual abilities, may significantly impact their abilities to interact with all aspects of assistive technology for success at home, school, work and recreation.

Screen Glare and Reflectivity

In automotive and aviation situations, discomforting, disabling, and blinding types of glare along with interfering reflectivity can be life threatening. These critical human factors must be anticipated and addressed. Also, glare and reflectivity are essential human factors considerations related to screens which most of us tend to overlook, pun intended. That is, we who have relatively normal-range mobility and sensory capabilities can adjust our body positions and angles, and/or those of our screen devices, to accommodate for, limit, or eliminate screen glare and reflectivity. This is not so for many individuals with disabilities, particularly those who by reason of age, limitations, or challenge cannot communicate to their support personnel that glare or reflections are limiting their viewing of a screen, and that they need their position or that of their AT device to be adapted. Sanders and McCormick (1993, pp. 533-547) offer an excellent discussion of glare and reflectivity. They state, "Glare is produced by brightness within the field of vision that is sufficiently greater than the luminance to which the eyes are adapted, so as to cause annoyance, discomfort, or loss in visual performance and visibility."

Sanders and McCormick (1993), Grant and Brophy (1994), Kantowitz and Sorkin (1983), and others further describe primary types of glare as direct or reflected. Direct glare derives from bright illumination sources (an unshaded window in full sunlight, a bright lamp, etc.) that the viewer can see within his or her field of view. Reflected glare may originate from these same types of sources but is light that has secondarily bounced off other reflective surfaces. These surfaces could include a CRT, LCD, or LED screen used on an AT device. Sanders and McCormick (1993) further categorize reflected glare as specular (from a highly polished, shiny surface), spread (from a pebbled or etched surface), diffuse (from matte-type, flat-finished surfaces), and compound (any or all these varieties of glare in combination). Another way of understanding glare and reflectivity is by their impacts on the person viewing them (FAA, 2021). Sanders and McCormick (1993) also summarize these types as follows:

Discomfort glare: Glare that is mildly annoying but does not interfere with a person's performance of visual tasks.

Disabling glare: Glare that interferes with performance of vision-based tasks and is uncomfortable for the viewer.

Blinding glare: Glare that derives from direct or reflected light, so intense that the viewer cannot see amid the glare. The viewer also may not be able to see other things well for some time after the source of the glare has been canceled. All are dangerous for drivers and pilots.

In most AT applications, instances of blinding glare are rare, although they might be faced by a user of powered mobility devices looking directly into headlights of an oncoming car at night, or by a user looking directly into sunlight outside or through a window while trying to work on a computer, or with other AT controls or devices. Though potentially able to have a serious impact on the technology user's performance, such occurrences of blinding glare in AT are probably less common than the other types.

More common in AT are the first two types of glare mentioned: discomfort glare and disability glare. Both can affect the visual performance of AT users, especially when that glare goes unreported by a user who cannot communicate well—and goes unnoticed by a teacher, clinician, or family member who may be unaware or uninformed of the potential hazard glare poses to optimal performance. Increasingly, instruction, remediation, and rehabilitation with children and adults who have a variety of special needs is being carried out via the medium of computer-aided instruction. Devices with various screens are involved in many of these programs, and of course are part of other AT devices as we have mentioned earlier. It may be an understatement to say that uncontrolled glare, direct and reflected, that is discomforting, even disabling to AT users, is present in many more instructional and therapy sessions than we know. The impact that any types of screen glare may have on assessments, evaluations, and interventions with special-needs clients and students, adults and children who may not know how or be able to communicate the situation, is of considerable clinical significance. Glare can cause on-screen images and information to appear distorted or partially altered, or to be not visually discernible at all. Asking students or clients to carry out screen-based tasks on which we will then evaluate and judge their capabilities or progress in intervention is a practice of questionable value if glare is not controlled.

222 | ADD HUMAN

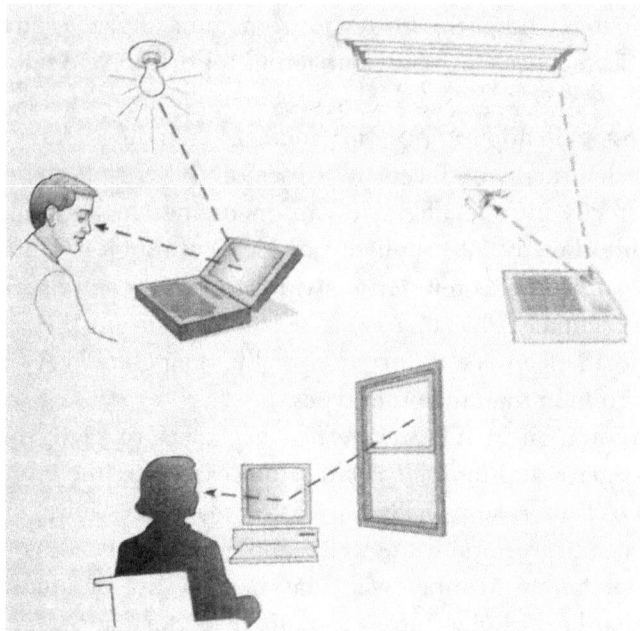

Positioning and angulation of screens on AT devices can be critical in reducing or eliminating major types of glare: discomfort, disabling, and blinding. If the screen display itself cannot be moved, then the viewer and/or the lighting source must be repositioned or shielded. Not all AT users are able to request, report, or accomplish this on their own. Family members and other supporting personnel must be alerted to glare from AT devices, and its potential to impact efficacy for users. Even clothing, furniture and room/cockpit colors and surface characteristics may affect reflectivity and glare in home, employment, driving and aviation settings.

Awareness of the problems glare can present is the first step. Taking concrete action to limit or eliminate discomfort and disabling glare in as many AT use situations as possible is the next. Some practical ways of anticipating and controlling for glare from screens will be presented and discussed. Fortunately, several effective, practical strategies and techniques do exist for detecting glare and minimizing or eliminating it. Based on Ostrom (1993), Grant and Brophy (1994), Sanders and McCormick (1993), Kantowitz and Sorkin (1983), Deming (1996),

Helander (2006), as well as our own years of clinical experience and numerous, frequent consultations with others who use AT and practice in the field of assistive technology, here is a listing of some considerations and tips that can be valuable in minimizing or eliminating screen glare for AT users:

Look. View the screen from the same height and angle that the AT user does or will. Get your own eyes at the same level, at the same distance, with the same lighting conditions, and at the same viewing angle as the student or client with whom you are working. What do you see on the screen? Do you see reflections or glare spots? Do they blank out or distort information or images on the screen? Do they tend to draw your interest or attention, and to distract you from the information displayed on the screen? If you see these types of reflections, it is probable that your AT client does too. If reflections or glare are distracting or even just noticeable to you, chances are they are distracting and annoying for your AT client, even if he or she has not said so. These disturbances in his or her ability to view the screen clearly almost certainly will be (or are) having a negative impact on his or her ability to interact visually with that screen in that position. Some changes need to be made.

Look again...and linger while you look. You cannot make accurate, useful observations in a brief glance from the AT user's point of view. Spend some time—several minutes at least—interacting with the screen in the same way and in the same position and location that you expect your student or client to do. Use the keyboard, the mouse, direct selection of on-screen "buttons," and so forth. Do enough of a simulation of your client's activity so that you truly are viewing the AT screen display device as he or she does. Observe different cursor and image positions on the screen, different brightness and contrast levels, flicker, focus, background and foreground colors, fonts, type styles, symbol patterns, and any other variables that AT users may encounter. Try to observe and determine if glare or reflectivity is adversely affecting your viewing of any of these other screen parameters. If any of these affect your viewing of the screen, it is safe to say that they affect your AT user's viewing of the screen as well. These factors must be dealt with in ways that will help your AT user.

Right angles of viewing to outside or strong light sources are best. Place the flat plane of all screen surfaces at 90-degree angles to strong sources of light. This will cause the angle of reflected light from the screen not to closely correspond to the AT user's angle of view (although it may create glare or reflections for an external viewer who is helping or looking on). Some of the powerful luminance sources to be considered may include bright lamps, track lighting, or outside windows into which the sun shines directly at different times of the day. In some cases, this 90-degree rule is easy to follow, especially for fixed video workstations in a room where the lighting environment can be controlled. In other cases involving AT devices with screens that are used in mobile, portable applications, it may not be a simple matter to control successfully for optimal illumination and screen angles. For upright screens such as the monitors used with desktop computers, the only readily feasible choice is to have the screen positioned facing at the user in a right angle to countertop and incoming light from windows. This works well if the monitor is viewed in rooms with light sources on a horizontal plane relative to the monitor or device screen, and with light sources that are not directly behind the viewer. In rooms with only strong overhead lighting, where the angle of the screen face is now parallel to that of the ceiling-mounted lights, glare may be a major problem. Most computer monitors have some built-in adjustment in their swivel base so that the screen can be tilted as much as 15 degrees up or down to help compensate for viewer preferences, glare, and reflections on the screen. This is generally not enough angulation, however, to compensate or correct for strong glare and reflection-producing lighting that breaks the 90-degree rule and that is immediately behind or to the side of the viewer. Other measures may have to be improvised and taken. Be alert.

. . .

Eliminate, block, or reduce strong sources of light. It often is not possible to move or reposition screens and the devices that they accompany. Then you need to deal with the offending sources of glare and reflectivity. Dark blinds can be placed over windows, and strong overhead lights or room lamps can be turned off or dimmed. As Ostrom (1993) cautions, however, in schools, libraries, or other public buildings or employment settings, one must check with fire safety engineers or officers to make certain that these lighting changes do not compromise fire safety for others in the building. Kantowitz and Sorkin (1983) and Sanders and McCormick (1993) recommend retaining at least 200 to 500 lux of ambient lighting in a room where CRT screens are being used to allow the viewers to look around the room. This allows them to readjust eye muscles as they focus on more distant targets—a necessary break from the eyestrain induced by the close work of computer access. Intense sunlight from windows without drapes or curtains can often be blocked just on the offending side by a temporary, portable shield constructed from a piece of cloth or cardboard suspended from a wall or ceiling. The shield can be taken down or folded up during parts of the day when the sun's glare through the window is not a problem. Similarly, temporary cardboard or fabric shields can be used to block offending light from permanent ceiling or wall fixtures that cannot be moved, turned off, or dimmed. In these cases, extreme care must be taken to avoid creating a fire hazard by putting such combustible shields too close to hot light fixtures. This is especially true with incandescent and halogen bulb lights. Cloth, cardboard, paper or other flammable materials must not be used where they can fall onto or move up against light fixtures. Serious fire hazards from such situations have been documented and must be avoided.

Protect the AT screen display itself from strong sources of light. Attachable glare screens made of cloth, plastic, or glass can be purchased for and mounted on many screen display devices. These provide reasonably effective filtering, light polarization by some of the glass or plastic units and protection from glare, although glare screens may not compensate for all strong sources of light coming directly at the screen. Screen brightness and contrast controls will have to be adjusted to compensate for the damping of screen luminance that occurs with all types of glare screens. Plastic and glass glare screens should be cleaned often and protected from scratching. Translucent sheer cloth glare screens require frequent regular cleaning because they readily collect dust that will block light as more dust accumulates in the weave of the cloth.

A glare hood is a piece of opaque plastic, metal, or other material that blocks light from reaching the screen or the viewer's eyes. Glare hoods can be mounted on the top of the screen so that they shade the screen from surrounding ambient light sources. Many communication devices and other tech with LCD screens can be fitted with a glare screen made and supplied by the manufacturer. These are opaque plastic shields that mount on the devices, and can be adjusted and angled by the viewer to block impinging light sources. Similar type glare hoods can be purchased or fabricated for use with other AT devices to shield screens from interfering light and reflections. Some have suggested making improvised glare hoods for CRT screen monitors out of cardboard boxes by cutting out the box and fitting it to the monitor.

A head-mounted, portable glare shield with appropriate cosmesis (a baseball cap from a favorite team or group) can help block glare in outdoor settings. This may even pertain with indoor settings that have strong overhead lighting which can discomfort the user or wash out screen images on a device, hence the cliché green eye shade used by accountants for years.

This should be avoided because such a glare hood can block air vent ports on the monitor and/or computer CPU, leading to overheating of the devices and resultant damage. Extreme overheating from use of homemade glare hoods can also lead to fire if the monitor (and CPU if it is covered, too) and its fan vent ports are blocked (Grant & Brophy, 1994; Ostrom, 1993).

Glare hoods may be most appropriate for and effective with some screens, both monochrome and color. These types of screens are particularly prone to being affected by glare and reflectivity. Fortunately, many of these screens are components of laptop, notebook, and other devices, and can be opened out, angled, and easily repositioned in relation to an AT user and surrounding light sources. These flexible, adaptable features of glare-prone screens are some compensations for their propensity to be affected adversely by ambient light.

In some cases, the viewing individual may be the direct, personal user of the glare filter or shield. We all have used sunglasses to reduce glare or our hand to block the sun from our eyes. These constitute portable, person-mounted glare filters and glare shields. A head-mounted opaque shield can be a workable solution for some AT viewers affected by strong direct overhead glare into their eyes, such as from the sun or from institutional lighting. The simplest solution to block direct glare in these cases may be a baseball cap from their favorite team, of course with a properly adjusted back band and front brim. A translucent, green-tinted bookkeeper's or croupier's plastic eye shield mounted on a headband are other examples of a person-mounted, portable glare shield. Both offer potentially effective blocking of direct discomforting or disabling glare, while retaining accepted, appropriate, even "cool" cosmesis.

---

100 words about *GLARE, REFLECTIVITY AND SAFTEY*

Discomforting, disabling and blinding glare can be deadly. Loss of a driver's or pilot's clear vision for brief moments at high speeds, or even while backing in/out from a parking lot or garage, can bring mishap and injury to humans, animals, vehicles and property. Ideas: Replace cracked windshields. Clean them. Drive slower with careful distancing in traffic. Use assistive polarized eyewear. Time your travels to avoid intense sun at bad angles. Reduce dashboard clutter. Visor

> **extensions. Pilots can face added glare risks. Even solar panel farms are glare hazards if too close to airports.** See www.safetybuiltin.com/glare *and* https://www.nhtsa.gov/sites/nhtsa.gov/files/811043.pdf **and** https://www.faa.gov/data_research/research/med...

Font Sizes, Types, Contrast, and Labeling are of particular concern with students, clients, and patients who are or are becoming literate. Interacting with letters, numbers, and other character data on any screen requires that the data at all locations on the screen be displayed in a maximally visible manner for the user. Although the American Optometric Association (1994) has suggested an eye-to-screen distance of 0.5 to 0.66 meters (20-26 inches) for computer users, secretaries, typists, and other persons using upright screens, this specified optimal distance is not always readily achievable for AT users who have special needs. Attention by teachers, clinicians, and family members to attaining this distance through careful addressing of seating and positioning concerns for AT users can be of great value in this pursuit. Coverage and discussion of these concerns is provided by McEwen and Lloyd (1990) and by Church and Glennen (1992), who offer more specifics about seating and positioning with students and clients.

As these authors have discussed, management of lordosis and kyphosis (spinal misalignment in sitting), foot, knee, and hip angles, and overall postural security can be successfully undertaken with most persons if proper expertise is brought to bear. Close cooperation with the OT, PT, rehabilitation engineer, and physiatrist can assure that optimal ways of seating and positioning the user are investigated, found, and relayed to others who can help implement them for and with a client. These variables must be considered in a collaborative team effort across the potentially great variety of lying, sitting, and standing postures a client may need throughout the day, and across the variety of screen viewing needs the client may encounter at home, school, work, and recreation. Attention to basic postural security issues for an AT user can help to idealize his or her eye-to-screen distance for this multitude of activities and situations. If the client cannot adjust his or her own posture, then, of course, the family, teachers, therapists, and other supportive care personnel must be alert to the need for helping to maintain this optimal 0.5- to 0.66-meter distance throughout the day. They also can help clients or students to be alert to and preserve it on their own, as they are able, throughout the variety of activities and positions they experience each day.

Font size is a human factor and variable in screen readability that can have a sizable impact on client success. The default setting for font size across type styles included with most word-processing programs for computers is often ten. A person with 20/20 visual acuity should be able to clearly see and use this font size from the optimal eye-to-screen distance of 20 to 26 inches over a workday. In addition, to enhance clarity and reduce eye strain and fatigue, many experienced computer users add the bold feature to make the font size stand out from the screen background. Increasing font sizes in this way aids many screen users in more clearly seeing their text and data on screen, especially over long periods of time, often four or more hours of continuous use. Similarly, the font size itself can be increased with most devices or software.

Some persons with mild to moderate cerebral palsy, for instance, have reported informally that when they try to maintain posture and focus on an upright CRT computer monitor, they prefer at least 14-, 16-, or 18- point bold font sizes because these make their interaction with the screen so much easier and less fatiguing (Gullerud, 1997; Johnson, 1995). Increasing font size and bolding the typestyle also can make reading of hard-copy printed information (on paper) easier for the AT user—and probably all others in the environment. Bolding and increasing font size takes up more space and reduces the number of characters that can be displayed on a screen or printed on a page. This reduction is of secondary concern, however, especially when it is more than compensated for by the increase in users' ease in seeing and reading of their work they are displaying on screen and printing to paper.

Size and boldness of characters and other information on screen displays is not always modifiable with all technology. Some augmentative/alternative communication and environmental control devices, powered mobility control displays, and radios or other gear may not allow changing character display components. Yet these portable, smaller, lighter devices may be more easily positioned in relation to the client's eyes, depending on how the device itself is mounted or positioned for access. Compensation therefore may be made for limited screen adjustments by the device being more movable and adjustable to the client's visual range needs; it may be moved closer to or farther from clients so they can see information on it more easily.

Other exosomatic (external to the body) solutions such as reading glasses, contact lenses, or screen-magnifying lenses can be considered for use with students and clients for whom use of these other nonelectronic technologies may be suitable. On many such devices with fixed-size, monochrome LCD screens, the "screen density" or "view angle" is readily adjusted with a small external wheel or slide-lever type of control. These controls should be experimented with as described above under the "Look" and "Look again" steps described under glare and reflectivity.

In brief, we must be certain to adjust the screen density or screen angle controls always from the client's point of view, not just ours. Ideally, these controls can and will be adjusted in all changing lighting and proxemic settings as the client moves throughout the day. Even more ideally, the client will have a method of controlling these factors on his or her own. It is important that appropriate clients receive incremental instruction and practice in doing so. Independent monitoring of these variables by students, clients, or patients is always most desirable if it can be achieved, with backup help and reminders provided by support personnel.

The font used in screen displays and other AT device readouts also can affect user viewing. In general, font styles that are simple and that rely primarily on straight, largely unadorned lines combined with smooth, simple curves are most easily read. Examples of these types of fonts include Arial, Arial MT Black, and New Times Roman. Each of these font styles is commonly available in most general computer word-processing software. Each has a minimum of extra curves, frills, and flourishes, and is elegant, easy to read yet simple in appearance.

By contrast, font styles that more closely resemble script writing, that are fancier and more complex, or that involve asymmetrical, unique character formations and combinations are harder to read.

They require a period for most viewers to decipher their nuances and for readers to become accustomed to their individual characteristics. Some examples of these more complex font styles include Benguiat Frisky, Lucida Casual, Old English, and Wingdings, among others.

Although these font styles may be highly suitable for some print applications, as in printing out greeting cards, casual letters, and simulations/approximations of handwriting for those clients or families to whom this is a concern (Angelo, Jones, & Kokoska, 1995; Angelo, Kokoska, & Jones, 1996; Leonard & Romanowski, 1995), overall they are less legible on both screen and on paper. Users, teachers, clinicians, and family members may wish to avoid these fonts, except for special occasions, and help AT users to become facile with the simpler, easier-to-read font styles for most adapted computer access tasks, schoolwork, and writing for employment purposes. Clarity and legibility in written form for characters on screen as well as on paper should be the overall aim of font selection.

Contrast of the font style with screen background colors and designs also must be considered for overall clarity. Most computers and word-processing software packages allow choice of screen background colors, and many also allow color selections for characters. Greater clarity for AT users and other viewers of the color or monochrome LCD or CRT screen displays will be had if the characters displayed are opposites from the background screen color. In other words, black letters on white (or light gray) background, white letters on black or darker blue background, and so forth. Colors of letters and numbers and the backgrounds on which they appear can be adjusted to users' and other viewers' tastes, while keeping in mind that maximal contrast between foreground and background will contribute to optimal legibility. Colors of characters and other on-screen information can play a particularly important role in helping make information distinct on maps, diagrams, or other graphic displays that contain lines, angles, and curves that may appear similar to type fonts. Displaying letters and numbers in a distinctly different color or various colors from background graphic material can aid the viewer in locating, seeing, reading, and interacting with the data displayed. Use of contrasting colors make the alphanumeric information stand out even amid considerable background clutter.

Labels and borders are additional effective methods that can help make alphanumeric characters and information more distinct on-screen displays. These techniques are useful on plain monocolor screen backgrounds and with visually cluttered or otherwise busy backgrounds. A label is a variable sized patch of color or gray shading that can be placed around numbers, individual words, or sections of text to highlight them. A label is generally of a consistent color that is subtle in shade to make the words or data contained within its borders stand out from the surrounding background.

Labels can be generated by many consumer software programs used in general computer applications, as well as in adapted computer access. Labels can be produced with many readouts. They are useful and common with most screen display types. Portable AT devices such as VOCAs that incorporate active color matrix LCD screens can make excellent use of labels as the surrounding highlighting behind static and dynamic symbols. The size and color of the label can be modified relative to the enclosed symbol, icon, or picture to make the label background obvious to find, and to make it stand out from the screen background. Labels can be moved

around the screen and positioned in places that optimize the user's ability to discriminate foreground from background information. More dynamic display AT devices are appearing on the market that incorporate LCD screens, with labeling used as an aid to be locating, viewing, and interacting with specific information on the screen.

Screen feedback regarding the user's entry of information also can provide important human factors cues to maximize successful use of an AT device. This basic feedback function of most software programs that run on screen display devices, whether a personal computer, desktop, laptop, tablet, phone, watch, or a dedicated VOCA, is often taken for granted. Nonetheless, it provides vital information about response of the device back to the AT user via sound beeps or chimes, light flashes or other visual cues, tactile responses such as key crunches or vibrations, and combinations of these, depending on the device and software.

Earlier we discussed feedback from switches and controls as providing important data back to users about how they are interacting with a control device. Similarly, at the larger level beyond the switch or control pass keys, feedback from the screen of an AT device can inform users that they are successfully entering information. Screen feedback can help indicate that the device is responding to input from keyboard, switches, and other controls as desired.

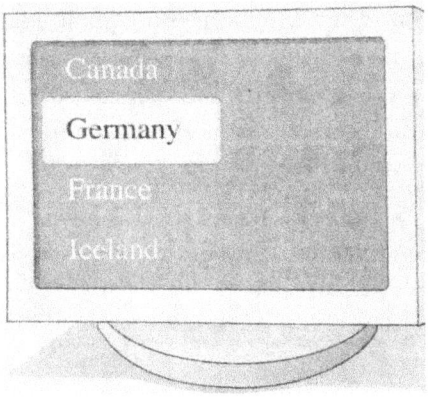

Use of contrasting-color, on-screen labels can help define and highlight text or graphic information. Labels and borders are commonly employed to assist users in locating and selecting important items on a screen.

Screen feedback can signal information regarding accuracy or errors of text and data entry. It can also signal keyboard and mouse cursor positions, page position, battery levels, incoming telephone or fax messages, or other information. Feedback from the screen might take the form of pop-up icons, dialog boxes, color shifts, stars, asterisks, screen blanking, or screen flashing. Audible signals may be paired with any of these based on changes to the screen and data or on-screen images displayed that have been caused either advertently or inadvertently by the user. Feedback in its simplest form may just tell users about which key, switch, or control was activated, and whether it was correct, and may indicate such obvious, commonplace things as end of a line, or side or bottom of page. Screen feedback such as a flashing light, attractive dot segment or animation, or a full screen flash may signal a ringing phone or doorbell for someone living with hearing impairment. Useful, easily activated adapted access features are built into most current computers and tablets to help make screen access and feedback more appropriate for many users living with diverse special needs and preferences (Proctor & Van Zandt, 2008).

Screen feedback is also important in telling users of peripheral functions that the dedicated computer or other microprocessor device may be running simultaneously with the display on screen. These peripheral items or other functions may include such devices as a printer, telephone, or fax modem, as well as battery charger cycle and levels, type of synthesized speech output, and so forth. Either through split-screen or partial-screen capabilities, or through changes that briefly and temporarily affect the whole screen, peripheral operations such as these can be monitored on the AT device screen while it serves another primary purpose such as communication, word processing, or environmental control. Overall, screen feedback allows for immediate monitoring and correction of conditions which the device needs to signal to the user.

Static versus dynamic displays of text, graphics, symbols, icons, and pictures on screen are the two major methods in which information may be portrayed on a screen display device. Static displays are common on CRT, LCD and LED screens and consist of information in text, symbol, picture, or other graphic form that is fixed in position, design, and content. That is, the version that appears on the screen (or touchscreen, or on the activation sites of a keyboard-operated device) must remain as it was originally set up and programmed, unless items are specifically reprogrammed to appear constantly different. Static displays of information on the screen (or on the activation-site array) of an AT device have several advantages. Primarily, they are fixed, stable, reliable indicators of meaning. Those users who learn, remember appearances and locations of, and then become reliant on static symbols and other static information displays can know that there are dependable landmarks and strategies around which they can plan and execute their interactions and activation responses with a given device. Romich (1994, 1997), Glennen and Decoste (1996), and others have stated that static, fixed arrays of information can help AT users to locate and use information displayed visually. This is due to the constancy and dependability of the appearance and position of the visual information. It is in the same place and looks the same all the time. Users do not have to hunt for or be distracted by visual information that changes appearance and location on the screen as they move among arrays.

Also, dynamic displays of data, graphics, pictures and other information are changeable and are found useful by many. These types of screen displays can be modified in size, color, content, and location on a screen display. The symbols, graphics, or pictures shown in dynamic displays also can be animated to some extent. Dynamic symbols can be made to move in appropriate, alternating patterns so that they draw the attention of the user, and to depict a concept such as "bend" or "jump" more accurately. Several picture and symbol systems on the market are examples of collections adapted and developed for dynamic display devices with this capability. This type of animation capability also can help to clarify the intent of a given symbol, picture, or icon, although some AT users and family members report that the rapid flashing, flickering on-screen activity of animated dynamic symbols on an AAC device can be distracting or annoying over time (Denning, 1996; Friel, 1997). Dynamic displays of information, particularly with portable devices are useful in assistive technology, with an increasing number of applications as they are developed further.

Although dynamic displays have many advantages, as mentioned, some have cautioned that the constantly changing nature of dynamic displays can be disorienting, confusing, and cognitively complex for some potential users (Glennen & Decoste, 1996). Some children or adults with special needs who by reason of age or disability may not yet have the cognitive, linguistic categorization or visual abilities to track the changing, dynamic array of symbols may find use of dynamic displays a daunting task. Tracking each changing symbol array, visually scanning and remembering it, and then linking it to previous or subsequent arrays as the screen display changes can involve visual, cognitive, memory, and motor requirements that may be still too challenging for some users (Former, 1997; Glennen & DeCoste, 1996; Romich, 1997).

100 words about *SCREENS VS SPECIAL ASSISTIVE LENSES: E.L.M. and LOUPES*

Eyeglasses are common precision assistive technologies. Other specialized lenses can rivel screens for image enhancements. Dermatologists perform *epiluminescence microscopy*, ELM, with small handheld lenses. ELM dermatoscopy uses focused, bright, polarized light to visualize skin structures below surface tissues, enhancing diagnostic accuracy. Dentists, surgeons, jewelers and others use high-powered Loupes, glasses with powerful, specially designed lenses to magnify and filter wavelengths of light. Loupes enlarge views and enhance color contrast of tissues or surfaces, allowing dentists to distinguish a cavity from a stain. Surgeons can better judge integrity of tissues, structures, and materials using Loupes. Jewelers and others use Loupes when precise visualization is difficult using unaided viewing methods. See https://madermatology.com/a-dermatologists and www.loupedirect.com/common-types-of-loupes-used-in-dfferent-industries

Proxemics, Seating and Positioning

Human factors that address users' proxemics, seating, and positioning in relation to the screen are also important. To a large extent, these factors are related to and intertwined with the previous two major areas we have discussed. The type of screen, as we have discussed, has much to do with how portable or movable the device it is incorporated with or used with can be. CRT screens, bulky, heavy and disadvantageous in many ways, are still in use in some places. They can be brighter and have vastly more viewing surface area than the other types of screens, but their size and heft may restrict their ease of positioning, portability, and mobility relative to a given user living with disabilities. LCD and

LED screens are much lighter, thinner, more easily moved and repositioned, and more portable. They can offer expansive viewing areas and generate brightness usable in diverse conditions, inside and outside of buildings. Once, some LCD and LED screens were more prone to having images "washed out" by direct and reflected glare, but that has been largely resolved.

All these factors above, plus others we have discussed before, affect how easily the screen itself can be adjusted or positioned appropriately for users. In this section, we will extend from the information covered already and look more closely at just the reverse: how screen characteristics might influence where and how the user should optimally be positioned or seated. Sometimes choices can be made—sometimes not. We will attempt to discuss some pertinent options and considerations for each type of screen, beginning with LED types.

LED screens have the advantages (and disadvantages) of often being small and of displaying a limited set of alphanumeric characters on some devices. LED readouts also have relatively high luminance ratios with most ambient light environments, so they can be exceptionally easy to see and read (whether they glow in red or blue light). This ease of viewing is true for persons with normal vision as well as, often, for persons with reduced visual acuity. AT devices that contain LED readouts as essential components include, as mentioned, several AAC, mobility, and other devices. Add to these the numerous watches, clocks, radios, TVs, thermometers, and other common devices that surround us in our homes, schools, and places of employment. Often, the LED device is so small and movable that enhancing a user's viewing of it simply involves moving it to a better position on a table or counter.

Glare and reflectivity typically are not problems in reading these screens because of the high luminance characteristics of LEDs and because the unit with the LED display can simply be tilted in the hand or moved slightly on the table or shelf to allow a better viewing angle. Also, LED devices such as wristwatches, remote control devices for TVs, CD players, and EC units can be secured onto bed frames or mobile supports on wheelchairs to bring them closer to the viewer. The optimal range for mounting of them for maximal visibility is something that can be explored through trial and error with a given student, client, or patient. A few minutes of holding the LED readout in different positions around the chair, bed, workstation, or other location can generally yield a placement decision as to where the device is most secure and where the user can see it best.

These adjustments are not difficult, but they are essential. If the readout on the phone or clock alarm must signal the user to get up on time for work or school, or to calibrate and assure timing for medications, therapies, or other important activities, then their ability to see it clearly is important. Having the display within optimal visual range is not a trivial matter.

Fortunately, the demands placed on user positioning or seating for viewing these types of screens are minimal because of the exceptional portability of devices that contain LCD and LED readouts, seldom the case with most applications of CRT screens. CRT screens by virtue of their weight and size require special considerations in relation to how much they can be moved; not all possibilities are safe or mechanically feasible. In applications that still must include CRT screens, it is preferable to have the bulky, heavy screen device positioned in a permanent location, and then reposition the viewer in relation to the screen. This may present some marked difficulties when the user is attempting to interact with a CRT screen while working from a supine or prone lying position in bed, or working across a large workstation table or counter.

A CRT screen computer monitor set on a table or cart at the foot of a bed or counter for direct-on viewing may only be solidly, safely mounted a body length or so (possibly one to two meters) from the user's eyes. Being able to see and focus on screen information at that distance can be exerting over long periods of time, even if fonts or image sizes are increased dramatically. Simply being able to tell where arrow or mouse cursors are on the screen becomes a difficult task for persons with normal vision at that distance, let alone reading and the other precise discriminations of text, graphics, symbols, icons, and images needed for schoolwork, employment, games, on-line communication, and other uses. Screen magnification via software, external screen lenses, or internal screen adjustments (80- vs. 40-colunm display, for instance) can of course be employed, but then the amount of visual data with which the user can interact per screen becomes reduced. The CRT screen can be moved to the bedside or side of a countertop workstation for closer viewing. This may be a viable solution in some cases, except that now muscle tension issues due to prolonged side looking may surface after a period of use.

Also, flicker of the CRT screen being viewed from a more side-on position (instead of directly in front) may lead to user annoyance, irritation, fatigue, and possibly even seizure activity (for those who are seizure-prone), which may be triggered by prolonged exposure to the flicker inherent in CRT screens. This may be exacerbated by looking at the screen from a side-viewing position over a long period of time. Gamma and X radiation are no longer major concerns for persons who use screens at least the recommended distance (0.5-0.66 meters) from the front or side of the CRT monitor. These used to be important considerations, but improvements in screen engineering have eliminated most of the potential for harm from spurious radiation.

To help solve a number of these potential difficulties, CRT screen monitors have been (and can still be) mounted on extendable arms that fold out or swing out to bring the screen closer to the user for direct viewing. With the use of these mobile supports, the CRT screen can be moved in closer to the ideal range (about 0.5-0.66 meters) from the viewer's eyes, as recommended by the American Optometric Association (1994). These supports, however, require permanent, secure mounting in a supporting beam, post, or wall to be safe for users and those around them. Successful use of these mobile supports also involves extra expense and time in selecting, ordering, cabling, and mounting—and then it all may have to be moved if the user changes rooms, residences, or worksites. Trade-offs are always present in AT; ideal configurations are difficult to achieve for any given AT user or system.

Coming as close to the ideal as possible for all other factors that must be considered in positioning users and visual displays are now LCD, LED, OLED and AMOLED screens. Their weight, size, and overall bulk are minimal, allowing for flexible applications across a variety of settings. Because of their light weight and trim profile, these screens do not have to be mounted in a fixed position-—that is, resting on a solid table, countertop, or cart. They can be moved, tilted, propped, even suspended in ways that are not readily feasible or safe with CRT screens. This flexibility makes them ideal for many applications, if safety and all types of screen glare are accounted for, including reflected, discomforting and disabling variations of glare from windows and lights, especially if users cannot reposition themselves. This factor, and the overall size of most modern screens (smaller than CRT screens) means that they are most useful if positioned relatively close to the user. This can be done with a variety of bed tables, carts, extendable arms, and other positioning and mounting devices that will safely hold a screen device. If needed, the screen itself can be detached from some tablet, notebook or laptop computers so that it can be suspended or mounted in many ways, close enough to the user, so that angle and screen density to reduce glare and allow maximum viewing are readily possible. The entire computer does not have to be mounted—the keyboard portion or add-on may be detached, separately cabled, and mounted for use in a variety of creative positions. Most separate screens allow horizontal mounting directly overhead (facing down) for a user who is lying on his or her back in bed or in a mobile support chair. Flat screens can also be tipped 90 degrees or any other angle and mounted sideways or on end. (Try that with a CRT screen!) This can help optimize viewing for clients who may be positioned lying on their side for AT access and computer or VOCA use during portions of the day. The capability for viewing from one's side may be especially important to some AT users at night. Communication, interaction and environmental control can indeed continue after dark.

---

100 words about *CLEANING PLASMA, LCD, LED, OLED AND AMOLED SCREENS*

Screens become dusty & dirty, collecting finger smudges & prints, even food/drink spills, making data difficult to view. During gaming, writing or working with numbers, we ignore screen grunge even as we strain to see details. Periodic cleaning brightens our work as well as our screens. With Plasma and all the other screens, care must be taken. A microfiber cloth for cleaning eyeglasses serves well, even removing tougher surface marks or stains if moistened a bit. Tissues, paper towels, your shirttails or your socks are

harsh and can leave lasting scratches on your screen. Spraying chemical solvents, cleaners or water directly onto your screen can also harm it permanently. *Adapted from Consumer Reports magazine, February 10, 2022.* www.CR.org

---

SUMMARY

Variables defined and discussed in this chapter affect successful use of consumer and assistive technologies. Human factors considerations with screens, displays, arrays, readouts and monitors are poignant as we list and describe them for persons living with challenges and disabilities. Cognitive, sensory, linguistic, physical, and time loads placed on all users by these factors can be heavy. There is no value in compounding these loads by adding distractions and barriers that can be avoided or compensated for by attention to basic human factors. CRT, LCD, LED, OLED AND AMOLED screens, properly selected, positioned and utilized, can prevent needless frustrating, discouraging obstacles in users' paths to technology success. Moreover, if essential human factors are addressed, screens can become reliable gateways into valuable lifetime technology use for consumers, students, clients, and patients of differing abilities.

REVIEW QUESTIONS

1. Name specific examples of Plasma, LCD, LED, OLED, AMOLED and CRT screen devices you have seen or used in your daily life. How did you or others use them? Specifically, in what settings? With what lighting? From what positions and distances?
2. What specific advantages and disadvantages can you identify for each type of screen, display or monitor you mentioned above in its intended application(s)? What adaptations, if any, do you have to make to increase the effectiveness of each? What wise human factors insights (or frustrating limitations) in the design of each device have you discovered?

3. Identify one or more assistive technology devices with which you yourself are familiar that include each of the types of screens discussed. What advantages and disadvantages of each of these screen displays have you observed during your students' or clients' (or your own) use of them? How did you deal with any difficulties encountered?
4. Can you make any generalizations about the physical size and strength assumptions for the "typical" tech user? How might they differ from users with exceptional abilities or needs? What interactions might vision, hearing, size and strength variables have in the practical, daily use of technologies by students, clients, and patients in different settings during the day? How might you address them?
5. Discuss how the numerous variables that we discussed that are related to displays of information may interact with an AT user's visual abilities, positioning, setting (inside, outside, etc.), and other factors. What do you think are the three most important variables related to display of information? Why?
6. Describe three types of screen glare and practical measures to prevent, limit, or eliminate each. How are these measures situation specific? How have you been affected by glare?
7. Overall, our efforts with screens used in AT should include simplifying, making selective, clarifying, and illuminating the information displayed. Describe practical measures you can take in your home, business, classroom, clinic, or other applied settings to achieve these goals for your consumers, students, clients, or patients.
8. What changes, improvements, or retention of current features would you suggest to designers and manufacturers of screens for general consumer and specialized AT devices? Why and how should they be accomplished?

8

TECHNOLOGY LEVELS, LITERACY AND LIFE SPAN

Technology is best when it brings people together.
Matt Mullenweg

The technology you use impresses no one. The experience you create with it is everything.
Sean Gerety

Technology is nothing. What's important is that you have a faith in people, that they are basically good and smart, and if you give them tools, they'll do wonderful things with them.
Steve Jobs

Assistive technology may include a wide range of users and options for complexity of tools and devices. AT and other technologies are often conceived of as necessarily "high tech," involving speech-input/output devices, powered wheelchairs, elaborate screen displays, or other expensive and complex electronic or mechanical devices. This does not have to be the case. We have mentioned the potential range of AT levels from no- and low-technology to high-technology in previous chapters and will further refine these concepts in this chapter. We will also propose the concept of technological literacy as it pertains to use of general consumer technology, and specifically to the use of assistive technologies by persons living with special needs, challenges and disabilities. This chapter will conclude with a discussion of some essential considerations in AT practice with users who at different stages in our human lifespan. Characteristics of learning and other essential human factors can change over different stages in our lives (Sanford, 2016; Shorrock, 2017; Spacey, 2021; Stanton, 2005; Wheatley, 2018).

LEVELS OF ASSISTIVE TECHNOLOGY

No-Tech AT Systems

When we refer to a "no-tech" approach in the field of AT, we are talking about harnessing existent body capabilities of the student or client without involving any external tool or device to accomplish some purpose to meet the challenges they are living with. For example, learning an adapted way of scooting or crawling up and down stairways without use of a lift device, crutches or walker could be a no-tech mobility solution for a young girl who may otherwise rely on AT for moving around her home and yard. The no-tech method must be learned and practiced to become a viable fix, but it requires no equipment or expense once it is mastered.

The term endosomatic (within or part of the body) could be used as a synonym for no-tech methods. Concrete examples of no-tech systems for communication include natural oral speech and speech approximations. Or, based on our statement above, better examples might include a man's use of an upward eye gaze for "yes" and a downward gaze for "no," or a woman's fingerspelling and manual signing with her hands instead of speaking. Examples could also include a child's simple forward head nod or sideways head shake to indicate affirmation or negation. No-tech systems for communication can also be formed by a user employing intoned vowels and vocalizations that, though not full speech, can still be effective conveyors of meaning for some persons in many communication contexts. Eye-gaze or hand-gesture Morse code, manual alphabets, or other codes, as described by Silverman (1995), King (1991), and others, can be effectively utilized for efficient communication when individuals move certain body parts in rule-governed patterns. No-tech methods of mobility could include learning to slide around the kitchen floor on one's bottom or learning a way to roll or sit to a better position from which to go up or down stairs. No-tech ECU might include learning to flip light switches or door handles with an elbow or foot to make the modifications needed, without using external electronic or mechanical devices.

These are just some of many examples of valid no-tech methods that can be developed and employed. Many other creative no-tech solutions across all areas of AT have undoubtedly been explored and created by users, families, and professionals. A compilation of just these would make an interesting book.

Note that in all the no-technology case examples cited, a "system" is formed by the user employing movement of one or more body parts in combination with only a set of rules or procedures to accomplish a definite purpose such as mobility, modifying their home or work environments, or communicating. These can be correctly referred to as "no-tech" methods: effective, assistive systems to accomplish purposes by the user without addition of an exosomatic (external to the body) device or tool. In summary, no-tech means "no external tool or device"; nothing other than the user's body is employed to accomplish the desired task. Next, we will look at "low-tech" AT, the next step up the hierarchy of assistive technology types.

Low-Tech AT Systems

Most AT systems are low-tech. Indeed, some AT users, their families, and professionals argue that the best, most durable, most reliable, and most cost-effective AT systems are often low-tech in nature. Low-tech (sometimes referred to as "lite-tech") systems involve a reliance by the user on rules and patterns movements coupled with a simple, common nonelectronic, nonmotorized device to accomplish a purpose. Examples of low-tech AT devices include a spoon with an enlarged handle to allow the user a better grip during eating for a person with severe arthritis or other limitations. A cane or crutch for walking, a mouthstick or headstick for typing on a keyboard, or a reach-extending "grabber" for grasping items on a high shelf are also examples of low-tech AT.

They are simple yet effective. Their mappings are natural, their affordances are evident, and their constraints are logical and obvious. Low-tech AT systems can serve as excellent starters for new AT users, reliable backup systems in the absence of more complex systems. Low-tech AT can often work well and reliably.

High-Tech AT Systems

Despite their transparency, simplicity, and usefulness, low-tech assistive technologies sometimes cannot accomplish the purposes desired by their users. The trusty pencil will never be a computer, even though both may be used for writing. Thus, high-technology tools and devices do have a significant, lasting place in AT. Persons living with severe limitations and disabilities need the extra assistance of electrical, electronic, mechanical, and/or hydraulic amplification of their natural abilities that only the more complex, more costly high-tech systems can provide. Examples of high-tech AT devices and systems may include wireless EC units, speech-outp3ut/input communication aids, powered wheelchairs and scooters, plus adapted-access computers, tablets and phones for writing, drawing, calculating, and telecommunicating.

With many of the current high-tech devices, we are looking at potential user-device systems that are not just augmenting or supporting existing user abilities. Often, high-tech devices may by more correctly viewed as providing alternative user-device systems for some individuals. In other words, they may substantially *replace* a person's limited or nonexistent abilities in mobility, communication, environmental control, play, or in any of the other component areas of AT.

High-tech tools and devices typically translate as higher-cost and higher-complexity, tending to also be higher on the opacity scale. Still, the more elaborate and powerful device may offer more potential to replace lost abilities if the user accepts the device and becomes skilled with it. Sometimes, the hidden trap of high-tech AT devices is focused on just that point: The device is capable of far more than the user is able or motivated to do with it. An unused, rejected or abandoned high-tech, high-cost AT device is as worthless to that user as no device at all.

TABLE 8-1. Levels of Assistive Technology: Practical Definitions and Examples

Levels	Characteristics	Examples
No-tech	Strategies, methods and/or techniques that rely primarily on user's ability to move or utilize various body parts Unaided or endosomatic Can be free, no expense Commonly transparent or translucent in use	Sign language; fingerspelling Knocking on a door Scooting up and down steps in modified way Eye or finger pointing Tapping or squeezing hand Vocalizing and speaking
Low-tech	Incorporates strategies, methods, and/or techniques, as above, with relatively simple materials and equipment commonly found in living and working environments Can rely on off-the-shelf, inexpensive consumer technologies Commonly simple; transparent or translucent in use	Handwriting with pen or pencil Modified eating utensils Picture or symbol communication boards Adjustable or reclining chair Elevated or adjustable countertops Basic wheelchair; stroller Modified knobs on faucets
High-tech	AT user integrated with complex, typically expensive electronic, mechanical, and/or hydraulic technologies to accomplish user's purposes Complex combinations of above technologies Commonly translucent or opaque in use	Sending a fax or e-mail Operating a speech-output computer Operating a motorized wheelchair Using an electric stair lift Modified controls and lifts for car or van Voice-operated ECU

The interactions of all the essential human factors we have discussed so far with no-, low-, and high-tech devices and their users are complex and probably infinite. The following scenarios composited from real consumers, students, clients, and patients we have worked with over the years will help illustrate some of these important points.

Scenario 1

"Mr. Halvorson" was 76 years old and had recently received a total laryngectomy operation; his entire larynx had been surgically removed because of laryngeal cancer. Natural voice and speech production were no longer possible for him. He was now medically stable and several weeks post-surgery. He strongly desired communication rehabilitation, of which several methods were explained by his SLP and tried by him in the clinic. Overall, the most appropriate (and the one he enthusiastically chose) was purchase and use of the Servox electronic artificial larynx (neck buzzer) unit. He quickly learned to use the device for alaryngeal speech, and appeared willing, even eager to use it for speaking in many situations. About three months later, during a follow-up visit to the SLP from whom he received therapy, Mr. Halvorson presented at the session using a plastic flip-erase pad from a variety store to communicate. He would quickly write or draw out his thoughts on the pad with a wooden stylus, then pull up the surface layer of plastic to erase and start again with new ideas. Mr. Halvorson reported that this write-flip-erase-write-again system worked well for him. He said that despite the cost of the Servox (around $600 U.S. at that time) and its relative ease of use for speech, he found over a period of trial use that he did not like the buzzing on his neck and in his hand—and that he and his wife hated him "sounding like a robot." He had put the expensive Servox away in a drawer and found that his use of the plastic dime-store flip-erase pad (89 cents U.S.) met most of his daily communication needs well since he was retired and just "hanging around" he stated. He liked using his eyes, face, hands, and body movements to help get the message through, and said that he did not feel as embarrassed or as much like the center of attention when he used the flip-erase pad—some important things the Servox, for all its complexity and cost, did not afford him. He joked that he could have bought a lot of flip pads for what the electronic larynx had cost.

Scenario 2

"Raul," a student from a small Central American country, related how in his village, a family had a young adult son with severe cerebral palsy whom they literally carried wherever they needed to go in the village. The parents were aging and found this method of mobility for him becoming too burdensome, even prohibitive. They sorely needed a wheelchair for him to help meet his mobility needs. Through an international aid agency, a manual push wheelchair was eventually secured and delivered to this man and his family. When they began to use it, they found that the frequent rains in their rural village in the forest area of their country caused the streets and walkways to be constantly muddy and too soft for the wheelchair to be used; it simply bogged down in the dirt with the weight of the son. Despite their delight and sincere (and often laborious and effortful) attempts to make the wheelchair work and to use it on these muddy streets for their son, the parents soon became discouraged because of its unsuitability for their village terrain.

Several months later, the man's father found an old Radio Flyer children's red wagon that had been left behind in the village by another family. He repaired a broken wheel on the wagon and discovered that he could pull his adult son quite well in the simple conveyance over a variety of surfaces. The four equal-sized, evenly weighted wheels and the front pulling handle of the wagon (which also afforded lifting of the front end) worked better for transporting his son than did the narrow, more unevenly weighted wheels of the wheelchair. Although this certainly was not their initial "technology plan" of assisted mobility for the young man with cerebral palsy, the simple red wagon worked better during most seasons on their muddy rutted village streets and walkways than did the more costly, complex, and ostensibly more appropriate standard wheelchair.

Scenario 3

"Jane" was a high-school graduate who had also completed over three years of technical-college education. She had severe spastic quadriplegia and severe dysarthria due to cerebral palsy. Despite this, she was a surprisingly fast typist (around 50 to 60 keystrokes per minute) using just her right index finger on a standard electric typewriter and QWERTY keyboard. Although she was a bright, intellectually capable woman who had many thoughts and ideas, her spoken communication was essentially limited to familiar communication partners who would take the time to understand her disfluent, slow, low-intensity speech. At age 38, Jane was still unemployed and was at last assessed and evaluated by professionals skilled in a range of AT for use of more sophisticated augmentative communication devices and systems. Her high levels of cognitive and language abilities, as well as her excellent receptive and expressive literacy skills honed over many years of formal education, made her a good candidate for a highly generative AAC system that included spelling, writing, and printing of information, as well as speech output. At the time of her evaluation, the most appropriate device for her was determined to be a popular commercial VOCA with synthesized female speech output. This device was made by a firm in the US, but is no longer produced.

Jane received the device via funding from Medical Assistance in record time—within about 30 days after the final evaluation. She began using it immediately to speak with her family and care providers. She also particularly liked and often used the paper-tape printout capability of the device for writing letters, greeting cards, and for leaving notes for her care staff and others. The voice output communication aid was an appropriate choice for her because it had features matching her skills and interests. It met many of her communication needs. While using it, Jane and the VOCA became a true human-machine system that accomplished her intended purposes in many settings.

As Jane used the device over a period of several years and became more interactive with communication partners in a variety of settings, she also matured and gained social skills and confidence. She began to realize that as she increasingly went to bars, restaurants, and concerts to listen to her favorite music and to have fun with her friends, her VOCA was not fully meeting her communication needs in these settings. Its speech output was not loud enough or sufficiently clear for her friends to hear, and the faint, small-font print on the LCD screen and the paper-tape output of the device were too slow and too hard to read to be useful to her and her listeners.

A major confounding factor in all of this was that Jane carried her VOCA in a backpack hooked to the back of her wheelchair. The device had to be pulled out by a helper and placed in front of her on her lap or on a nearby table for her to use it. These were often unstable, ungainly positions for her to access the device efficiently and safely. This also required the assistance of someone else to dig out the VOCA from her pack, and further reduced Jane's capability for successful independent use of her AT for communication, especially in the low-illumination, high-noise settings found in bars, restaurants, and concerts. Jane was also particularly pleased with her recently acquired color-coordinated, sleek wheelchair that matched with her favorite wardrobe colors and complexion. When it was suggested that she needed a mobile mounting system on her chair that would allow her to always have her VOCA more accessible, she vehemently resisted. She emphatically maintained that even though her VOCA would then be in a position that was far easier for her to access, the bright chrome pipes, rods, and big black swivel joints of the mounting device would not look right on her new wheelchair— that it would all make her appear "too handicapped." She did not want to be seen in public that way, she said, and in addition would

not agree to such a mount being put on and ruining her attractive, cool new purple chair. Cosmesis of her AT devices was a central, essential issue for Jane.

Some creative problem solving with Jane resulted in her developing a small, low-tech, direct-selection-based "pointing board" of words, phrases, letters, and numbers that she carried with her on the wheelchair at the side of her leg. When she needed to clarify, correct, or elaborate a spoken message in a loud or dim environment, she could bring out the plastic-covered pages on her own and point out information to repair the communication breakdown. When she was in a more conducive environment, she used her low-tech system to request that her VOCA be placed in front of her for speaking and/or writing. Over time, the increased sociability and independence that use of this low-tech pointing board gave Jane caused her to consider how having her VOCA available on a portable mount might additionally enhance her abilities to communicate in fuller, more precise, more elaborate ways in many more communication contexts. As of our last discussion, Jane was considering whether she wants the mounting system but was leaning toward use of it more than before. She may be willing to trade a bit of cosmetic appeal for system functionality.

These are just a few of the relevant clinical anecdotes that could be shared. Based on our examples and information discussed here, what human factors and trends in assistive technology must still be addressed? It is evident from reading the AT catalogs, attending AT conferences, and even just following news reports in the popular press that AT devices, tools, and systems are proliferating. The application of human creativity to the solution of specialized AT needs for disabled persons is truly impressive and humbling; it is inspiring to watch human creative powers harnessed for such worthy purposes. But what about human factors in AT? What essential human factors considerations are being made with these new products? Are they addressed, or are they ignored as not being relevant or important? And what still must be addressed by consumers and manufacturers? Let us turn our attention to what the state of the art is—and then to what it could and should be.

Beukelmann (1990, 1995), Church and Glennen (1992), Glennen and DeCoste (1996), and others have discussed how we have moved recently from the initial amazement and discovery phase of AT, into a more subdued, long-term, public-policy-oriented view of assistive technology. In other words, rehabilitation and education professionals, and to some extent consumers in many of the industrialized nations, are now past the initial blush of excitement and enthusiasm over the power that AT can confer on individuals with special needs and severe disabilities. AT works. We know it works. Its power is great and can change lives. But now that AT has become more commonplace, professionals and consumers alike recognize that there must be life after device acquisition. Just getting the nifty AT device to the user is only a small part of the picture. The user-device honeymoon ends after a few months or so, and should then lead into a successful, lasting, growing marriage of the user with his or her assistive technology.

As Beukelmann and others addressed, what perhaps matters more now in the United States and in many other Western countries is public policy regarding acquisition of AT for appropriate, deserving users, and then, the surrounding, real-world issues of purchase, funding, fair pricing, device design, and device acceptance by the user and by society at large, as well as device repair, replacement, revision, and/or upgrading to newer versions over time. (Do you still use that fountain pen or VHS player or eight-track player you had decades ago?) Each of these issues includes complex factors related to societal, psychosocial, and socioeconomic variables. These issues also relate to national and world market forces, social trends, customs, religious and philosophical underpinnings of any given society or subculture, and numerous other variables.

The complexity of each of these—and then the interactions between and among any or all of them—is beyond the scope and purpose of this current text. They are mentioned here in passing to alert the reader to these incredibly important and complex factors, and the potential of their critical mutual interplay. All these societal factors in assistive technology deserve more complete research and more extensive coverage and discussion later in a larger volume.

As indicated by the more frequent appearance of the topics of human factors and device rejection in the rehabilitation and education fields, and by the emergence of this book and others like it on the market in to the 2000s, recognition of essential human factors in assistive technology is occurring on a broader scale. We have been enamored of the glitz and gee-whiz factor in AT for during this initial stage of the field's development, and we clinicians, educators, and makers of AT have overlooked an important part of the puzzle: These devices are intended to become integral parts of people's lives. If tools and devices are to do this, especially with persons who may not otherwise be readily able to protest or adjust factors they find undesirable, attention must be given to all the human factors we have discussed, and others we have not covered. Market forces shape general consumer products so that they are functional, compatible, easy to use, and pleasing for the purchaser. Products that do not work well or that do not look good to users simply disappear from the market in the open competition for consumers' acceptance and purchase of them.

The AT market, however, is much more limited in size. With some exceptions, such as glasses and contact lenses for mass supply, expensive assistive technologies are often more pertinent for personas living with moderate to severe disabilities, a lower-incidence, lower-visibility population. Until the last few decades, persons with severe limitations and challenges have been largely overlooked and under-recognized by mainstream societies in developed countries, and ignored or unknown in other less developed nations. In the United States, passage of Public Law 101-336, the Americans with Disabilities Act (ADA) of 1990 and its subsequent interpretations, along other similar and supporting legislation in this country and others, have made disabled persons a more definable, recognized minority. Nonetheless, severe disability is still infrequent across an entire nation's population. Encounters with disabled persons, let alone insightful relationships with them, and true understanding of their need for human factors considerations in specialized technologies, are rare for most others who do not have experiences with challenging or limiting conditions.

Therefore, design and manufacture of AT devices and equipment do not appear (yet) to have been fully influenced by consumer reaction to cosmesis, durability, logical constraints, natural mappings, and other essential human factors. Not enough users have yet reacted (or been able to react) to design deficiencies that result from failure to address human factors for these specialized technology users. Innovation, change, and adaptation to some of the finer user considerations, such as cosmesis requirements, switch and control sensitivity and selectivity, and device affordance, have been left to the whim of the designer and manufacturer, rather than matched to the needs of the potential or actual AT user. If few remark about AT, or even realize that they could complain or critique the exciting new AT devices they have sought so hard and for so long, then manufacturers continue to make and sell products without refinements that could increase device efficacy, efficiency, and acceptance. Well-meaning, learned professionals in the AT field continue to believe in, acquire, and try to use certain set technologies with our clients because these tools and devices, and the designs embodied in them, are what we have available and are whet we know ... and so on. The cycle continues without interruption until attention is called to deficiencies or refinements needed in the product's inclusion of essential human factors.

At times, failure with device experiences and learned helplessness for AT users can result along the way. These can be perpetuated despite our best efforts to help through assistive technology and supportive interventions. This book and other media can help to alert and sensitize users, families, and professionals to the human factors considerations they need and must insist on in the devices they use and that are recommended, as these human factors considerations are often missing. Resulting changes in AT consumer needs, desires, and purchasing patterns must be made known to designers and manufacturers, and can influence their future creation and distribution of assistive products. This cycle of consumer influence on the AT marketplace is not thoroughly researched, although to users, families and clinicians it seems slower and less responsive than the cycle of consumer influence on product design and longevity found in the design and life span of most general consumer products.

The central points of this book surface again here. Technologies must be transparent as possible, and devices should match with the humans' needs and characteristics—not the reverse. Useful, successful devices and tools must be designed to meet the needs of the user; the user should not have to accommodate unnecessarily to the demands of the device. These simple premises seem so obvious. Yet the incidence of tools, devices, and technologies (assistive and otherwise) that are too hard to use or do not mesh with the consumer's desire or capabilities seems high. Whether device rejection is based on cosmesis, functionality, or other factors, it is still rejection. A consumer device, or a more specialized, expensive AT device that is not used, is of no value no matter how expensive or hard-fought its acquisition was for the user.

One of the trends observed in AT recently is to make mobility devices, communication aids, environmental control units, and other assistive devices more pleasing to look at—to improve overall cosmesis of AT. Until fairly recently, it seemed that many (possibly most) AT devices and equipment items appeared "institutional." That is, they appeared as if they were made for rugged mass use, with little consideration of personal preferences, style, or personal expression embodied in the appearance of the device. Since a variety of color choices has become included, even expected, in most general consumer products since the mid-1970s or so, an increasing number of AT-related products have, decades later, also become available in colors other than boring chrome metal, or beige and gray plastic that once seemed synonymous with AT and special-needs users.

When discussing crutches, one user we know has referred to their typical color as "depressing gray." The same could be said for many such enabling products over past years. They are boring in appearance and look like a "handicapped device," as some have said. The bright, varied colors now of many communication devices, switches, and other AT devices and accessories are just a few examples of the responses to requests by AT users, families, and professionals. They have wanted and got more variation and attractiveness designed and engineered into the products they rely on through so many circumstances in their daily lives. Why should AT users and those around them not have choices?

Are your pens all the same color or type? Are your dinner dishes only white China? Are your eyeglasses only made of thick black plastic? Of course not. Manufacturers are increasingly finding ways to retain ruggedness and functionality in the AT gear they design and make, while still including the more pleasant outward product appearance that we have come to expect in general consumer devices and technologies.

A quick perusal of AT equipment catalogs reveals helmets, switches, bikes, wheelchairs, scooters, mounting devices, canes, prosthetics and most other AT devices increasingly available in a range of pleasant, attractive colors. This range allows the user to match the appearance of their tech with personal preferences, and to coordinate with the image of self that they wish to portray to the world. Although Henry Ford reportedly said in the 1920s that you can have any color of car from his company "as long as it is black," the fact is that now most consumers want and expect choices, even with helpful AT devices and equipment that used to be just depressing gray.

Until recently, the lack of choices in color and appearance of AT devices meant that a lot of items looked bleak, institutional, and like a "depressing aid for sick people," as a parent of another of our clients phrased it. Today's growing options in colors and in more pleasant external appearance of assistive technologies, tools, and devices should only serve to enhance user satisfaction and acceptance of AT in terms of the human factors of affordance and cosmesis. AT can more easily be viewed by users and others as accessories and parts of users' wardrobes, just as we accept glasses, pocket and wristwatches, and the related bands, straps, chains, or cords that may accompany these assistive devices. Magazines and web sites feature specific section on fashion and appearance for AT users, with some companies focusing just on designing and producing clothing designs, fashions and accessories for AT users and others with special clothing needs. These sources should be followed and consulted as fashions change and develop for AT users. "Dressing for success" should be a choice for all consumers.

Other readily evident areas in which design and manufacturing capabilities as well as user expectations have progressed, especially for electronic products for general consumers and for AT users, are size and weight. Phones, tablets, computers, televisions, radios, telephones and other electronic consumer entertainment devices generally began their existence over past years as devices so large and heavy as to be essentially immobile. In their original forms, they were mostly impractical for any portable application by consumers of average size and strength. Our consumer trends have been toward smaller, lighter, more portable and transportable versions of all these popular devices. The response of the purchasing public favoring portability and compactness has been powerful, as smaller devices to be easily carried or worn continue to proliferate.

This same evolution in size, weight, and portability is evident with AT devices. For example, the original electronic AAC devices were heavy, large devices that covered an entire wheelchair lap tray. They were difficult to lift, mount, and move for many users and their families. These behemoths of electronic AAC were followed in the early to mid-1980s by smaller, more efficient devices from many manufacturers. Some of these devices were still heavy, but at least were equipped with carrying handles and/or shoulder straps, making true moment-to-moment portability and use possible for many users. These 12- to 16-pound transition-era devices also could be mounted on wheelchairs or other mobility devices via sturdy mounts made with chrome tubing and hinging or swivel devices. Even though the mounting devices themselves were not necessarily streamlined or always particularly attractive, they did make the VOCA devices more portable, allowing them to become more fully integrated into all aspects of the daily lives of AAC users.

The advent of much smaller dedicated devices, and the flexible versatility of smart phones and tablets, and slim, lightweight laptop computers herald the acceptance and expectation of lighter, smaller, more practical AAC devices for high levels of portability. Just as we have seen with other consumer gear, this trend is growing with crutches, wheelchairs, communication aids, and other areas of AT. Physical sizes and weights of portable AT devices for communication can vary considerably, although the trend is surely to smaller, lighter devices. However, as devices become more compact, resolution and selectivity of activation sites can become significant issues for many users. Smaller does not always translate as easier to use or better even if it means greater portability.

Designers and manufacturers continue trying to meet the cosmesis and portability/size desires and needs of consumers of assistive technologies, just as they have tried to meet these needs for consumers of more general popular technology. As cosmesis of AT devices improves, more AT devices may be seen in public. More AT devices and equipment are visible in frequent open use as they become increasingly common and accepted by the public. The more they are accepted, the less stigma is attached to them. As AT devices and equipment become more accepted as the right tools for the right jobs, the more the affordance of them becomes apparent and accepted as appropriate: "You can't talk? Oh, you should use one of those cool speech-output tablets I've seen," or "You can't walk? Well you need one of those fancy wheelchairs I've seen race on TV."

Although these hypothetical responses may be simplistic, they reflect at least a general awareness of AT and its potential usefulness. The more a tool, device, or other item is used in public by others, the more it seems to become accepted for use in public among others. As with eyeglasses, the affordances of AT may eventually become so automatic and obvious that little thought or stigma is assigned to the helping technologies any user may adopt. Assistive technology presence and purposes will be considered more natural for all of us as we require it.

Notably, mobility aids such as wheelchairs, scooters, and stairway lifts also have become more streamlined and more attractive. The common institutional- looking chrome and gray plastic wheelchair has been supplemented now by choices of many colors, and sleek, faster, more efficient, and cooler looking designs. Similarly, powered wheelchairs and scooters can now incorporate color-coordinated enamel paint, upholstery, and safety straps to match apparel and surroundings. These refinements are making these devices appear more appropriate for shopping, school and office, and for socialization sports events, races, and Paralympics-type competitions across many exciting, reinforcing sports and games. These refined assistive technologies are becoming integrated into and accepted as normal parts of activities for home, work, school and recreation.

Just as automobiles and bicycles are fully accepted mobility aids on which the general population depends for daily use, and just as the telephone, cell phone, notebook computer, and other devices are fully accepted assistive communication devices used by most general consumers of technology, so too can specialized AT become accepted. The end-task purposes of AT devices across all areas of assistive technology are no different than those of their general-consumer technology counterparts. At their simplest, they are devices that help humans to move, see, hear, communicate, control their environment, play and recreate, and otherwise move the world to their advantage. As AT becomes more compact, more attractively and acceptably designed, and more frequently used, the cycle of acceptance can be enhanced. Things around us that are commonplace tend to become "invisible." Our hope and the intent of this book is to help motivate research so that product changes, and acceptance among the professional and general populace that will promote this growing invisibility for assistive technologies across all areas. Good things are happening, yet considerable progress toward universal design is still needed across many areas of technology.

This ideal of universal design and full integration of assistive technologies into the everyday experience repertoire and acceptance of the public seems lofty, perhaps impossible, but it has already occurred in several areas of technology use. The best examples, as we noted, are eyeglasses and corrective lenses. These are probably the oldest and the most widespread assistive technologies, used by over half of the U.S. population, and with similar incidence in other developed nations. The acceptance of eyeglasses and corrective lenses has become essentially universal and has been enhanced as designers and manufacturers have addressed consumer demands for cosmesis considerations. Glasses and lenses can be designed for school, play, and work, and for dressy and casual occasions. They are an omnipresent yet essentially unnoticed assistive technology. Hearing aids and other increasingly common assisted listening technologies now seem to be following in this pattern of growing general acceptance through attention to miniaturization, costs and enhanced cosmesis.

As another example, the typewriter, now word-processing computer, has become a universally accepted device. Its mere presence on a desk affords work productivity. Few would question that. Yet, according to Diamond (1997), the original concept of what we now know as the typewriter was developed to aid the letter-writing abilities of a blind Italian woman around 1808. This prototypical keyboard "typewriting" device, developed by Pellegrino Turri, was designed with keys laid out in a linear sequential fashion, much like those on a piano. This design and device eventually evolved into the typewriter, now the computer keyboard, which have become the standard mass technology for assisting written productivity in offices, schools, and homes for over a hundred years. The typewriter has now been largely, although not totally, replaced by the computer and electronic word-processing technologies in many settings. These more complex electronic devices are still based on the same central idea of turning keystrokes into words, rather than relying on handwriting.

Our common wordsmithing devices still retain the nearly universally accepted and used convention of the QWERTY keyboard, although many would argue that this configuration of characters is, at least for some users, ungainly, outmoded, or inefficient (Anson, 1990; Diamond, 1997; Kantowitz & Sorkin, 1983; Lloyd, Fuller, & Arvidson, 1997). As ungainly as the typewriter first appeared, and despite the initial reluctance of many to accept personal typing of information as being as valid as personal handwriting, it has become universally accepted for how it enhances productivity (Diamond, 1997). The typewriter and its progeny in many sizes, forms and complexities now and once so controversial, has become unremarkably commonplace.

It is our hope and belief, based on these and other successful integrations of assistive technologies into the mainstream of product use and acceptance, that (1) most specialized assistive technologies will evolve into unobtrusive, largely "invisible" devices that blend into daily technology uses by the general populace in most countries, and (2) societal acceptance of the use of a variety of tools, devices, and products of all types to meet users' needs will grow to the level of complacency we see now with telephone, cars, television, and other devices. Seldom do most users think of the often-incredible complexity of these devices and how well they truly serve their purpose. We users of everyday helping technologies tend to think only of the end products: the conversations, discussions, fun, recreation, and work productivity that result when we use these technologies. The device is generally overlooked in the pursuit of what it allows the user to do. We all fall for his tendency. Good... Once unique, arcane technologies are blending into our daily lives.

We are arguing, in essence, that when this same apathy and lack of excitement about common technologies can also more readily occur with the more specialized assistive devices that now still tend to mark, set apart, or otherwise stigmatize a user with special needs, it will be all the better. Sometimes, complacency is a good thing. It may at least be a benchmark of increased societal acceptance of assistive technologies into the broader and deeper mainstream of the many devices that people use to help accomplish a variety of common daily purposes. True integration of AT and users may be attained only when blasé becomes the best description of general societal response to them both. Novels, movies, life stories and real examples of success with AT will continue to educate and change us across cultures and nations as we keep on telling them to all.

TOWARD TECHNOLOGICAL LITERACY

Literacy is a key concept in our discussion. It can be variously defined. Its simplest and most common definition includes terminology such as "the ability to read and write." Other definitions include mention of discerning meaning from letters and symbols, and expressively producing written or drawn symbols, pictures, or iconic depictions that have semantic intent. In general, literacy has been understood to deal with the primarily visual skills of working with, manipulating, and processing letters, characters, words, and symbols to use them to convey meaning. Literacy understood in this light is surely an important component of human behavior, but perhaps this is too limited a definition.

Montgomery (1995) proposed an expanded, more inclusive definition of literacy that may have greater significance in the rehabilitation and special-education fields. She stated that literacy includes "all forms of speaking, listening, writing, reading, and thinking." Her expanded view of literacy challenges us to consider: Might literacy skills indeed involve the use of a rule system (language) to interact and convey meaning with others in a variety of ways? Could literacy include absorbing ideas through hearing or vision, and then speaking or writing analytically about them? Could literacy mean not just reading, writing, or speaking about a topic, but also using what has been absorbed from the sound and light to think about the topic? Should a composite definition of literacy, then, include the combined use of these other inputs to think about other topics and to create new ideas, synthesize them, elaborate on them, and critique them?

Our perspective on this is emphatically affirmative. Literacy, when defined as the aggregate of listening, speaking, writing, and reading, and then using all these components to provide the tools for thinking, becomes the focal point of much of education, special education, and rehabilitation. From an assistive technology perspective, much of the reason we use AT is to enhance or help replace one of those areas of literacy, speaking, listening, reading, or writing. Whether we are employing AAC, adapted computer access, assistive hearing or seeing devices, or most of the other areas of assistive technology, it could be said that part of what these modified systems help the user to do is expand literacy skills in one or more of the areas Montgomery describes. From glasses to hearing aids to speech-input or speech-output computers, all these high- tech devices and many low-tech ones can work together to increase the potential literacy capabilities of the AT user.

Thinking is perhaps the most essential component of literacy (Montgomery, 1995). Cognitive capabilities are enhanced by improved capabilities in the other four areas of literacy. Being able to speak and listen and to read and write about ideas gives a person the tools with which to be able to think and with which to think more complex thoughts. As humans can converse, share ideas, and interact through oral language and spoken communication skills, the more they can refine and expand on their ability to think. That is, the more a person can experience, share, and communicate, the more he or she will be able to remember, associate, critique, analyze, and synthesize ideas about that to which they have listened, seen, and spoken. Much of early language development and the early grade levels of formal schooling are devoted to just these tasks (Nelson, 1991; Shames, Wiig, & Secord, 1994; Van Riper & Erickson, 1996). Young children learn to listen and speak to gain and share information with which to think about increasingly complex and abstract ideas. Across all languages and cultures, children gain highly functional oral language skills within their first five to seven years of life. They rapidly become skilled in listening, speaking, interacting, sharing the thoughts of others, and developing thinking skills of their own. A 5- to 7-year-old, normally developing communicator may well have essentially adult linguistic competence and performance in his or her native language, as noted by Chomsky (1968), Van Riper and Erickson (1996), and others. Many of the component areas of AT, as already mentioned, can be of value in assisting growth, development, and ability levels of these speech- and hearing-based forms of communication.

TABLE 8-2 Some Components of Literacy

Hearing and Listening

Vocalizing and Speaking

Visualizing and Reading

Configuring, Writing and Calculating

and then using all the tools above for further

Thinking, Strategizing, Planning, Creating, Communicating, Coping, Learning...

In many societies, these oral language skills serve as the foundations for visual and graphic-based communication abilities which allow even more complex thinking. Reading and writing allow us again to gain access to and to express ideas, thoughts, and information—and to preserve these in forms that can be shared with others at times other than the immediate moment. We need to write down our thoughts and ideas and then to read our own writing or that of others to be able to think deeper, more complex thoughts.

As a practical example of this, divide 8 by 4 "in your head." Not too difficult, right? You know the algorithm and can "do it in your head." We can handle that type of problem almost reflexively. Now try dividing 8 by 7. Still mostly simple numbers, but you may have to write this problem down on paper or use another common assistive device (a calculator) to help you think through the problem and get the exact answer. Feeling confident? Try dividing 83,792 by 4,523. Even though this is still arithmetic with whole numbers, you will now almost surely have to use an assisting device to help with (i.e., write out) your thinking process. The long-division algorithm that you will use to solve this problem remains the same, but the details of the arithmetic have now become so numerous that few people can solve this type of problem, do this type of thinking, without using exosomatic aids. This is just a small example of the many ways in which technology can give extra leverage to our literacy abilities.

The points of these exercises are simply these; We can think about many things using our oral language skills based on speech and hearing. To think truly complex, multifaceted thoughts, however, and to be able to put them down in forms that we can critique, share, and preserve for later review, we must be able to write them out—and then be able to read back what we have written. Unless the information we are thinking about is quite simple and concrete, we almost always need some method of writing out and then reading back the information we are dealing with. Sometimes pencil and paper will suffice; sometimes the thinking task requires a complex, high- tech microprocessor-based device, such as a calculator or computer. In short, we may have to use a range of technology to think a range of deeper thoughts.

Whether we are using a pencil or a supercomputer to solve these types of problems or for a school assignment or a report at work, we are relying on technology to help us think. It is evident that devices, both simple and complex, can enhance our literacy skills. They not only allow us to see and preserve our thinking so that we can transfer some of the cognitive load from our cortex onto the paper or screen, but they also help us to organize our thinking—to see "where we are going" with our thoughts, and then to use the symbols of language and mathematics to connect with the literacy abilities of others who may wish to share our ideas—or we theirs. These literacy skills grow as we use them. Literacy experiences that are carefully guided throughout preschool, elementary school, secondary education, and postsecondary education allow us to deal with concrete and abstract information at higher levels than if we could only listen and talk about them. It would be exceedingly difficult to simply dictate this book or a lengthy term paper without the revising and correcting functions that viewing of the printed text allows. The sum of literacy allows us to use all our specific skills of speaking, listening, writing, and reading to give us things to think about. As we learn to organize and expand our thinking abilities through using these tools, we have more to think about in both quantity and complexity. The more we can think about, the more complexities we can express through speaking and writing...and the cycle goes on. Typically, the more we engage in literacy skills, the more skilled in literacy we become.

Assistive technology can help fill the gaps that some persons have in the component areas of literacy we have discussed. Although some of the specialized AT tools and devices may be different from what the rest of the population uses, AT is essential to literacy for persons with special needs and disabilities. Adapted devices—and ready access to them—for thinking about problems like those described here, and for the many other complex cognitive tasks allowed by and required for interaction through language, mathematics, and graphics, are essential for higher literacy skills. In the studies by Angelo, Jones, and Kokoska (1995) and Angelo, Kokoska, and Jones (1996), one of their findings was that some parents expressed greater interest in their AT/AAC-using children being able to have methods for adapted writing than for augmentative or alternative speech output. Findings from these parents express what we intuitively know: To succeed in school and at work, we must be able to write. The ability to write down our thoughts enhances our ability to think with more complexity.

So often our interactions focus on the spoken and listening portions of literacy, perhaps because these are the immediate developmental and social interaction concerns that we first encounter with other humans. That is, speaking and listening seem to be foremost in our considerations of interactions with others; we engage in speech and hearing interactions with other humans generally before we find out how well they can read and write. Nonetheless, in societies where formal education is important, writing and reading abilities eventually do become an important part of the "mix" of literacy abilities treasured by that society. Personal independence and success at many future educational, vocational, and social pursuits can relate directly to overall literacy skills across all five areas we have discussed. Those who have disabilities or limitations, or who cannot otherwise engage successfully in the variety and complex interplay of literacy skills expected in society, would appear to be at substantial jeopardy for reduced achievement, independence, and life satisfaction.

In addition to more standard conceptions of literacy discussed above, we submit that another term must be coined: technological literacy. **Technological literacy**, as defined here, refers to a person's (or, on a larger scale, a society's) recognition and knowledge of these four concepts:

1. **Devices and tools exist that can extend, amplify, and assist human abilities across a wide range of activities and intents.**
2. **These devices and tools can be readily found and acquired for one's use.**
3. **Use of these devices and tools can be learned or figured out based largely on one's prior experience with and knowledge of similar technologies.**
4. **Patterns of usage retained from previous experiences can help users accept, learn, and become skilled with new technologies.**

Technological or device literacy is based on our understanding of the previously discussed elements of general literacy. It is also derived and expanded from the current term computer literacy. Although this is a term derived primarily from popular use, a more detailed, formal definition of computer literacy would likely include several of the elements defined here as they specifically relate to computers and software. How capable one is in approaching and using a computer to write, work, play, and think may all be realistic components of computer literacy. This concept can be a platform from which to launch our more expansive concept of technological literacy.

In practical terms, technological literacy means that people in any given culture develop tools and devices to assist in their work and play. They become accustomed to and rely on these devices to accomplish the purposes desired. From a human factors perspective, true systems of user, device, and purpose are created and accepted. The systems that work well are reinforced by their success and further use. Experience with these tools and devices establishes a pool of learning and a technological experience repertoire among people in a culture that becomes the basis for other users to become aware of the existence and value of devices—and, later, for skilled users to adopt successfully new, different, perhaps more complex technologies as they become available in a culture, and so on.

In ways like how we tend naturally to develop skills with speaking and listening to aid in our interaction, play, and work, and then can develop reading, writing and other graphic literacy skills to enable our cognitive processes, so too can skills be developed with using simple to complex devices to aid our thinking, our play, learning, and work. Certain tools and devices become common in any given culture; competency (or "literacy") with them is expected. We witness this with devices ranging from knives, forks, and spoons, to sewing machines, hammers, DVRs, smart speakers and phones, and computers in a technological society. In other contemporary but developing cultures, such as the original, traditional cultures of the Hmong people, literacy with rabbit and bird snares, fishing nets, and cloth-making looms may still be highly valued and frequently used (Vue, 1995). What you become skilled—literate—at using depends greatly on what you are expected to become skilled at using in your native culture and country.

But what about the members of a culture who cannot use the tools and devices that are common in others' experiences? Where and how does technological or device literacy develop for persons who cannot have access to or learn to develop skills with the array of simple to complex devices with which competency is expected for others in their culture? For some persons in all societies who may be living with cognitive, sensory, linguistic, motoric or other special needs and limitations, exclusion from even beginning-level contact with many of the devices of their culture is the rule. The experience repertoire that seems so common among most members of a society as they grow and interact with tools and devices may not be shared by all. Personal disabilities, special needs, and limitations can disrupt the likelihood that someone will experience learn about the world through using tools. How can you experiment with devices and tools, and learn to become skilled with them, when you cannot manipulate them? The answer of course is that you cannot—and that you then are further excluded from developing even the basic level of technological literacy with simple devices that should become the springboard to eventual literacy with more complex devices.

A brief inventory of the common items, tools, devices, and technologies with which any of us interacts each day, from birth through senior adulthood, yields a myriad of things with which we typically become skilled. If you could not see them or move them or intuit the method of their use, how would you use them? What is the magical set of movements to tie shoes? How do you dial a phone? How do you coordinate cutting and eating with a knife and fork? How do you type? This list goes on and on. This may sound simplistic. Yet the premise is profound: In any given culture, we can use the more complex tools and devices we come to rely on because we first learned to use the simpler ones. Our literacy skills with more complex technologies emerge out of our earlier learning with less complex, seemingly transparent, obvious devices. If this early learning and experience base is absent or limited, the developmental transition to literacy with more complex, translucent and opaque technologies is jeopardized.

This is precisely the situation for persons living with significant disabilities at an early age. Their chances for becoming as technologically literate as others of their age group may become markedly reduced. The natural generalization and skill growth process from device to device to device is short-circuited early in their development (Kantowitz & Sorkin, 1983; Musselwhite & St. Louis, 1988; Sanders & McCormick, 1993). Based on these premises, our aims in diagnoses and interventions for meeting student, client, and patient needs through assistive technology would be oriented toward determining their current technological literacy levels. Then activities and learning involvements for them can be designed and targeted to increase their general device literacy—as well as device literacy specifically related to assistive and consumer technology use. Ultimately, this skill streaming attempts to create levels of technological literacy that will allow special-needs users to approximate the technological literacy levels of others in the culture of similar age. These life-learning experiences and lessons in technology must involve all component areas of AT as relevant to a given consumer. They must also include the collaborative practice of educators, therapists, clinicians, families, and others who can broaden and refine these interventions tailored to foster growth of an individual's technological literacy.

Exactly what these interventions are across each of the areas of assistive technology, and how best to integrate and deliver these to consumers of all ages, are topics that continue to require substantial research and methodical reporting of clinical experiences. Creative, efficacious practice on the parts of all who compose the AT team is required—we still have much to learn in these areas. We can be guided, however, by the knowledge that technological literacy with the entire range of relevant AT devices for an individual must derive from incrementally graduated, successful experiences with a variety of devices. Experiences and development of competencies with transparent to eventually translucent and opaque technologies must be included, with overall intervention design focusing purposeful efforts to create experience repertoires that will extend as far as possible across all levels of complexity.

ASSISTIVE TECHNOLOGY DIAGNOSIS AND INTERVENTION

To determine appropriate interventions with AT, a thorough, organized diagnostic approach must be utilized. We will review and describe a general approach used in AT that was originally promulgated by the Pennsylvania Assistive Device Center (PADC), also known as Penn-Tech. This approach has been adopted by many others. In this approach, as originally described by Haney and Kangas (1994; Stanton, 2005;), the diagnostic process is composed of three distinct yet integrated steps: assessment, evaluation, and diagnosis. We will elaborate on each.

Assessment

In the PennTech model, assessment translates as "collecting data." In the assessment phase, examiners must maintain a neutral mindset while data are collected about a child's or adult's history, AT needs and use characteristics, and other relevant areas concerning the client's personal, medical, educational, and psychoeducational background. Potential AT users are observed interacting with technologies and with people in a variety of real-world settings. Information on personal history and history of AT use, if any, is collected and recorded in an objective manner. No attempt to adjust, judge, or intervene is made during the assessment phase.

Evaluation

When data across the person's many areas of abilities and disabilities are collected, then informed judgments based on these assessment data can be made. This phase is properly referred to as evaluation. During the evaluation phase, value judgments are made about the data assembled. Do the data collected from the assessment phase (across many areas of behavior and performance) suggest that the user is within normal range for his or her age, size, and other characteristics? Or do data indicate that the person being evaluated performs above or below levels that might be expected of others in his or her cohort?

While the assessment phase is intended to be nonjudgmental and focused on the objective collection of data, the evaluation phase is focused on making value judgments—on determining where the characteristics of a person lie in reference to others of the same age, gender, and so on. In AT applications, we also are concerned about how these abilities stratify with reference to characteristics of specialized tools and devices. These may relate to control site and activation site matches, resolution, sensitivity, selectivity, mapping, affordances, and other essential human factors that may pertain to a potential AT user.

Diagnosis

The final phase of the diagnostic process is the diagnosis itself. Contrary to what we may have come to believe about diagnosis from a medical model perspective, diagnosis in AT, as well as in much of special education, general education, and rehabilitation, often translates differently. Diagnosis in these fields means the separating of the client's strengths and weaknesses (i.e., strong areas and not-so-strong areas) as they relate to possible use of and interaction with devices. The diagnosis phase of this process allows us, based on the previous careful collection of objective data and then the careful judgments we have made about that data, to describe the AT user's profile of strengths and weaknesses. It further allows professionals to determine what device and technology features, or characteristics may match with the user's patterns of strengths and weaknesses across all areas of ability and disability. It allows them to plan appropriate teaching interventions and device selections based on this feature-matching process, and to document baseline performance levels so that progress can be measured in valid and reliable ways for a given client. The diagnosis phase also allows us to make appropriate referrals to other professionals for other types of assessment and evaluation, and to seek further refinement of the diagnosis from others who may help the student or client meet important needs that we find cannot be addressed through interventions with AT. These areas may include dental, medical, behavioral, psychiatric, nutritional, or other needs that can influence or impinge on a person's successful use of AT.

Careful and thorough completion of the diagnostic process in AT allows for appropriate post diagnostic planning for outcomes, expectations, and activities for special-needs consumers who may be candidates for assistive technology use. First off, candidacy for AT can be decided more accurately by completion of the whole process. The coupling of the full diagnostic process with a trial-run phase, in which devices across all pertinent areas of AT for a given consumer are tried out in the real world, is a vital element of this process. Data are kept during this real-world experiment in an ongoing extension of the assessment phase to affirm or deny earlier judgments made in the evaluation phase. The beauty of this process is that, ideally, it can be a self-correcting, revisable process wherein early decisions and selections of appropriate AT can be borne out in observations of positive, successful AT experiences by the client. Or, if need be, these temporarily configured AT devices can be revised, altered, or adjusted if prior device decisions are found to be lacking in efficacy in actual applications in the daily life of the client. As final decisions are made from the diagnostic and trial-run phases, appropriate scaled interventions can be planned.

Intervention

Although we tend to use intervention as the umbrella term for what happens after the diagnostic phase, this portion of our interaction with a special-needs consumer can be more accurately defined in more precise, detailed increments. Intervention—or treatment, or therapy, or whatever else we as professionals in the AT-related fields do to help consumers of our services—can be divided into several relevant components. These include interventions that are intended to remediate, compensate, manage, or be "on behalf of" an AT user in relation to his or her educational, rehabilitation, and/or assistive technology needs. We will define, describe, and expand on each of these terms.

Intervention is the broad, umbrella term referring to the general scope of what we as professionals can do actively to help improve the quality of life for a student, client, or patient. Efforts to seek improvement of abilities through a variety of means across all areas of human performance and focused on increasing the likelihood for a person's successful involvement in school, employment, family, and the other central pursuits of life, can be considered intervention in the broadest sense. Synonyms for intervention may include treatment, therapy, training, and instruction. You may also think of others.

Palliative treatments or interventions tend to be comfort measures. Perhaps much (most?) of medical treatment given in cases of severe, chronic, or terminal illness can be called palliative in nature. These types of interventions help the patient to feel better and to cope better with the condition. From a technology perspective, perhaps the provision of calming music or visual effects, such as is available in a multisensory room (Lendle & Maro, 1996), or simply from an audio source, could be called palliative if these might help a child or adult to be comforted, consoled, or buoyed in spirit. No change in the person's abilities or technology use is necessarily targeted or expected—just their enjoyment of, comfort with, or improved attitude from the use of technology to make life seem better for them for a while.

At the opposite pole of interventions and their effects, iatrogenic interventions or treatments are ones that, while treating one disorder, cause another to occur. Laryngectomy is perhaps the most dramatic example of iatrogenic intervention. Removing the patient's cancerous larynx prolongs his or her life—but at the cost of introducing another severe disorder, which may require AT as a partial solution. Removal of the larynx means loss of it for phonation, effort closure, protection of the airway from foreign bodies, and for warming and humidifying of inhaled air. In this example, the dramatic yet effective treatment for one condition (laryngectomy for laryngeal cancer to save the patient's life) causes other dramatic, life-changing conditions (loss of voice and other related important laryngeal functions). In other situations, the cure may be as bad as or worse than the condition it was intended to help. Assistive technologies may help only in part.

The most held view of what special education, rehabilitation, and medical and dental interventions do is to remediate a problem area for a student, client, or patient. Remediation of disease and disability means fully "curing" the problem; to help the student, client, or patient improve to the point that, whatever the difficulty, it is erased, removed, taken care of completely. Certainly many medical, educational, and rehabilitative interventions can indeed accomplish full remediation. Many treatments allow improvement to the point of functional "zero difference" with peers, or with one's former self before onset of the affecting condition—a complete cure. We see this with many infectious diseases, some reading difficulties, some speech and language problems, and limb use difficulties experienced by some patients. Overall, however, the concept of cure probably fits much better with the medical or dental fields than it does with the fields of special education, rehabilitation, or the restoration of lost or impaired complex human functions in general. A philosophical and operational trap into which many new professionals in the therapy and education fields fall into is adopting a mindset of "total cure" or complete remediation as the only valid intervention goals. Their goals for students and clients and their personal expectations for practice are that persons in their charge will experience complete remediation of difficulties with which they present. When this total cure does not happen, the professional loses confidence and may become demoralized. Professional or personal burnout can readily result. Could this also be considered an essential "human factor" for professionals in AT? The same impact can be felt by families and students, clients, or patients themselves. Lofty expectations of remediation or cure in rehabilitation and special education may be better replaced with expectations for the value of palliative care, plus functional, if not perfect, compensation for impaired abilities.

Compensation, in fact, may be the focus of most of our efforts in AT. Abilities in the component areas of mobility, environmental control and manipulation, communication, seeing, hearing, or other areas that are not developing as they might be expected to for a child, or for an older person to whom these abilities have been lost due to accident or illness, may be compensated for by the technologies we have been discussing. Compensation means using something in place of, or to take over the function of, natural, endosomatic abilities that cannot be remediated. Sometimes compensation is also valid with abilities in process of being remediated (and for which prognosis is good) but for which a functional substitute is needed immediately.

AAC is an excellent example of compensation. A patient who is severely dysarthric or who has lost motor speech abilities completely may be helped greatly by an AAC device that can compensate for loss of oral speech. Adapted computer access can help compensate for lost writing abilities; powered mobility for lost walking and running skills, and so forth. Creative, acceptable, functional compensations for limitations and disabilities are the essence of assistive technology. Successful compensation is what the AT field is about.

In addition to compensation, remediation, and other direct types of intervention, management is sometimes the wisest strategy. Management implies that direct treatment, cure, or compensation may not be possible. The person involved, and his or her family or others, can be taught methods to use at home, in school, or on the job to minimize the impact of a disabling or limiting condition. For example, urinary incontinence in some cases is handled better by teaching the patient or family to make sure that the patient uses the bathroom at prescribed, set, frequent intervals throughout the day and night, rather than by catheterizing the patient and having her or him wear a leg bag for urine collection. Another example could include cutting food into smaller pieces and arranging them on the plate in an organized manner to help a person bite, chew, and swallow. Simple management strategies may afford more dignity and greater levels of acceptable cosmesis than some other more direct treatment approaches.

Management techniques seem particularly valuable in cases where the prognosis for positive change or cure of a condition is limited, but where the person involved and those around him or her are willing to learn and implement helping, managing strategies. Many management strategies are essentially palliative in nature. Management with AT commonly includes such seemingly mundane activities as device recharging, care and repair, adjustment, lubrication, and other daily skills that can help always keep compensatory devices and equipment effective and serviceable. Teaching and improving the user's and family's skills in these important day-to-day areas of AT can assist in managing the technologies they depend on; AT that does not work is no better than no technology at all.

Palliative intervention seeks to provide relief from discomfort of all types so the recipient can live better and easier in the present moment. It commonly includes helping the recipient make the best of each day they have remaining for however long that might be. Palliation focusses on efforts to assist the client or patient in coping daily with pain, stress, anxiety, agitation, and other symptoms that detract from their current quality of life. Care and comfort, rather than cure, are the objectives of palliative interventions.

Another category of intervention that is receiving more acceptance has been described as "on behalf of" care—in other words, doing things on behalf of students, clients, or patients to assist them without direct treatment (Ross Thomson, 1988). This can be of substantial benefit. For instance, an OT may train teachers and classroom aides in adapted pencil-gripping techniques that they can try with some of their children with low motor skills. The OT could be said to be engaging in intervention activities "on behalf of" those children, even though she may not be treating them directly. A PT who teaches a husband how to transfer his wife safely from her wheelchair to bed or chair and back is working on behalf of the patient even though he (the PT) may not be working directly with the woman. An SLP who consults with food preparation personnel at a nursing home or school lunch program on textures of foods needed for a client with dysphagia would be an example of intervening "on behalf of" this client or patient, even though the clinician is not working directly with him or her.

TABLE 8-3 Basic Categories of Intervention

Remediation

Compensation

Management

Palliation

"On Behalf Of..."

Combinations of All Above

An understanding of intervention types and purposes is essential to successful practice in AT, as in any field. Knowing the range and scope of interventions that can be offered and the limitations presented by each is. important in ethical, skilled practice by professionals of all types. Recognizing and understanding the range of interventions and their purposes is also important for AT users and their families. Realizing that care more than cure, or compensation more than remediation, are often the more valid and expected goals for use of AT may help alleviate some false expectations and help keep hopes for AT success within reasonable bounds. While we in AT-related fields of practice tend to be eternal optimists, hoping for the best for our consumers, it is also true that our assistive devices are not magic wands that can restore what has been permanently taken away. Assistive technology is never a simple panacea that we merely take out of the box and watch as it meets our clients' challenges. Helping our consumers and their families to understand that compensation and daily management may be the most important, overarching goals of AT acquisition and use is essential to ethical practice. Knowing that this is the case, may increase user and family acceptance, reduce guilt for remaining limitations and imperfections in AT use, and help clarify intervention goals for all involved.

TECHNOLOGY ISSUES ACROSS OUR LIFE SPAN

Life-span issues in relation to how people learn and how they use tools, devices, and technologies of all types, including AT, are receiving attention. For some time we have followed a homogeneous, "one size fits all" approach to teaching consumers of all ages relative to tool and device use. This has also applied to the design of special devices that we expect persons with special needs to become skilled in using. Increasingly, we are finding that how persons (whether disabled or not) of different ages view and approach technology may vary considerably. Some of this increasingly evident variation may have to do with device-related experience repertoires—but much may not.

People at different stages in the life span have different characteristics in how they learn, as described by Brookfield (1991) and others. We tend to recognize this intuitively, but our intuitions do not always guide our practice, nor are they necessarily supported yet by empirical research, which is still, itself, in its infancy. As Beukelman (1990) summarized, this may be especially true in how people learn to use technologies that are new to them. The difference in technology learning across the human life span is a fascinating and complex topic. An introduction to it is valuable for students, clients, and patients who may come to rely on AT, as well as for the families and professionals who work with them. We will summarize some important aspects of lifelong learning characteristics pertinent to technology. Knowledge in this area is, unfortunately, not yet vast. Much more research into what we do not know, plus refinement of what we do know, is vital.

Early Childhood

From birth to around age 3 to 4 years, tool and technology use by humans is marked by exploration and play; the child views tools, devices, and technologies as toys. It has been said that (young) children "learn to move and move to learn" (Arnheim, Auxter, & Crowe, 1977; Karp, 1987; Sherrill, 1977). This also can apply to learning to move things. Although this axiom may well apply to human learning at all stages in the life span, it is apparent that movement among and exploration of different devices, tools, toys, and technologies are critical to children's cognitive, linguistic, and motoric growth.

Although universal conclusions about how young children interact exclusively with the range of assistive technologies are not possible at this time, some general statements about what we do know can be made. Typically, very young children (babies and toddlers) tend to interact with toys, devices, and other items in a multisensory manner.

That is, they use mouth, tongue, face, fingers, hands, feet, and perhaps other body parts and surfaces, as possible, to taste, feel, and move objects, as well as to sense temperature, texture, movement, and other aspects of items. Their control sites (the specific areas of their body that they use to operate or interact with a device) tend to be the most varied as compared to all other age groups. (How many adults in your work setting try to push a switch with their tongue? They could; most people don't.) It is common, in fact expected, for an infant or toddler to explore an item first by putting it into his or her mouth. This area is the most sensitive and responsive monitoring system of young children, and they use it frequently to interact with items in the world around them. This propensity to "mouth" things can, of course, also present significant safety hazards for babies through normally developing 3-year-olds. This is common knowledge among parents, teachers, and childcare workers, and has been the foundation for policy changes related to toys and other small items sold and distributed for children by toy stores, fast food restaurants, care centers and schools. Bassock, 2019; Bell, 2019; King, D. 2022; Winkles, L. 2022, and many others have researched and/or commented on this topic.

When it comes to interaction with enabling technologies at this age, the likelihood is that young children will use parts of their bodies that older humans may not. It is also likely that young children may make more use of their whole bodies to interact with devices. This is due in part to developing differentiation in motor skills of the young child, evidenced by a tendency to use wider, larger control sites rather than more differentiated, smaller, finer options. That is, young children will use the whole palm of a hand instead of just one finger to push a button, and so forth.

Young children's learning of technology seems to be characterized by an approach attitude of eager play and open exploration. The extensive literature on child development and play should be consulted for greater elaboration. At this early stage of human life, consequences, either physical or social, for failure to use a device in the "correct" way do not seem to influence children's learning and exploring as much as with older persons. Ignorance of the stigma of failure can, for a while at least, be bliss.

Childhood and Early Teen Years

Elementary school-age children through children in their teen years exhibit considerable control site refinement. Most children of this age in industrialized societies have become skilled in using fingers and hands as their primary control site for switches, as well as for mechanical and hydraulic controls. Indeed, our world is configured as a hands-on world, from early childhood onward. Most of what we will touch, manipulate, control, or operate from then on during our lives in most industrialized societies will be done with our hands. There are not many alternatives among general consumer technologies. Pedals on a tricycle or bike, or foot controls on a car, are about the only frequent foot-based exceptions to the predominant hands-fingers orientation that is designed into virtually all other devices.

Incidentally, this is not necessarily universal. This focus may be different across other cultures and would be an interesting topic to explore further. That is, do humans in all cultures naturally tend toward finger and hand use as their main control sites for most tasks of living? Or do some cultures markedly depart from the seemingly obvious, universal affordances of hands and fingers? Some examples of this may include, but are surely not limited to, the use of feet in some Mediterranean cultures for crushing grapes, the Savate foot-fighting method of France, empty-hand karate in Okinawa, or the practice of carrying coins in the ear rather than the hand as reported by some persons from the Philippine Islands. Other such exceptions to the hands-only rules of industrialized society may well exist and would make fascinating study that may have relevance to AT use and learning across ages and cultures. In other words, hands may not always or naturally be our best control sites; they are not the only human options for interacting with devices. We may have other viable options we are overlooking because of constraints of our culture.

Nonetheless, industrialized societies such as the United States, Canada, and other countries seem to base technology use largely on devices requiring finger and hand controls. This predominance of hand- and finger-oriented design for devices, as well as the dominant hand/finger-oriented cultural mindset, can affect users who do not have or do not develop skill with those control sites. This cultural orientation to almost exclusive design for hand and finger use may have relevance to AT that we do not fully appreciate nor understand. In any case, children between 4 and around 15 to 19 years of age (early childhood into the late teen years) tend to be eager to learn about devices. They are becoming skilled users of their hands and fingers and are enthusiastic about exploring switches, controls, and the outcomes of their activations. Furthermore, they often willingly accept, expect, and tolerate direct instruction by an overseeing authority figure, someone who is there to teach them but who may also critically point out their mistakes and correct their developing skills.

Children of these ages want to learn more about how to use and interact with technology successfully. They still look to surrounding adults, as well as older and/or more skilled children to help them learn, to correct their mistakes, and to provide a model for further learning with technology. During these years, many technology skills with computers, keyboards, joysticks, and other controls are learned from competition and interaction with peers in experiences that focus on games and recreational uses of technology. This is also true for skilled manipulation and enjoyment of other noncomputer technologies and devices such as skateboards, skis, bikes, and cars—and many other items that may not operate on electricity.

The bottom line is that children of these ages, like the previous group mentioned, are typically willing to try and willing to fail with technology, and they are willing to try again after instruction or correction from others. Their motivation to continue to develop skill with the technology, whatever it might be, overcomes the cognitive, linguistic, physical, and time loads they may experience. You are encouraged to relate this to our earlier discussion and interpretation of Baker's basic ergonomic equation (1986) that summarizes motivation versus effort in AT use.

Further, Beukelman (1990) has stated that this explorative age group also likes to learn with "their hands on the keyboard"; they have little regard for the written instruction manual. They will make errors and mistakes "in front of anyone" without being daunted or becoming discouraged. Technology appears to hold only the promise of fun for them—^not the portent of failure or embarrassment. Those negative aspects of relating to technology seem to arise later in our life span.

100 words about *NEOTENY*

Cuteness can count in tech success. Designers, cartoonists, babies & kittens exhibit how neoteny attracts. Cute is our judgement as we view certain toys, people & animals. Neotenous features appeal. We want to engage, cuddle, have rapport with a baby, toy, or animal when they exhibit a larger head-to-body ratio; smaller mouth; larger, wider-spaced eyes lower on their face; smaller, softer nose; shorter fingers; and rounder, cuddly body. Combinations of these features engage us, evoking our responses of caring, teaching, protection. Neotenous features in cute toys, pets, art & designs make us want to buy, adopt & interact. Think pandas vs cobras. Both are unique. One beckons us to hug and pet it. Neotenous features in young and older individuals can offer selective survival advantages.

    See www.zmescience.com and https://en.wikipedia.org/wiki/Neoteny

Young Adults and Middle-Aged Adults

This age span is the largest, from approximately age 20 to around age 65-75. Correspondingly, we tend to see the greatest variability in learning styles during this developmental period of middlescence. Young adults emerging currently out of a high-tech video and online gaming, and extensive social media backgrounds may tend to be fearless in their approach to computer-based technologies. They have learned in a hands-on manner and are commonly eager and confident in their approach to further exploration and learning of these related microprocessor-based technologies. How aging may affect the technology use and skills of this cohort of young adults, and how their approach to and use of new technology evolves, are yet to be seen. It is hoped that they will retain the enthusiasm and confidence with technology that many of them have brought from their techno childhood.

This current generation of older children and teens has the first (and perhaps the best yet) chance to age with technological experience and confidence compared to any preceding generation. As they move into the workplace and experience the transition from technology that is used primarily for fun to technology that is used increasingly for work applications, it will be interesting to observe the evolution of their learning styles, technological literacy, and approach tendencies to technology, especially new technology.

On the other hand, this general age grouping of people includes middle-aged adults, now approximately 40 to 75 years old, who did not grow up as saturated with computer or video games or with extensive exposure to computer-based or other electronic technology. At least, they probably did not have as close or intense an involvement with electronic technologies as do the current cohort of young adults. In general, this age group of persons tends to be familiar with and frequently use some electronic, mechanical, or hydraulic technologies in employment settings for work-based applications. They are supervised, evaluated, and rated on their work accuracy and productivity by their supervisors and employers. In brief, their livelihoods depend on how well they use technology. Technology for them has probably become more of a way to get the work done well, rather than a way to pursue recreation and exploration.

These are, of course, huge generalizations, but they may have considerable face validity. Many middle-aged adults are familiar and comfortable with technology. But for the many others who are not, use of technology can be highly threatening, even disconcerting. Beukelman (1990), Brookfield

and Schauer (1997), and Sanford, 2016), among others, have described several characteristics of middle-aged adult learners of technology. These learners are commonly highly goal directed. That is, they want to learn specific applications of technologies to apply to specific tasks or goals right now. These goals pertain to tasks related to their employment productivity or to specific personal needs such as tax preparation, financial record keeping, generating and tracking holiday greeting letters, archiving family information, and so on. Middle-aged adult learners typically prefer to learn in private or with one trusted companion or co-worker; they do not want to be placed in public view while they are learning new devices. They desperately wish to avoid learning new skills and developing competence with technology in situations where they might be openly critiqued. They do not wish to be observed by clients, students, family members, or nontrusted peers or supervisors while struggling with new technology, and while making "dumb" mistakes with devices that are new to them. Preserving dignity and professionalism is essential to most learners in this cohort. The stage of learning technology for fun has often passed for this age group; they wish to receive actual instruction focused on the work they need to accomplish with the technology.

Valuable instructional approaches with this age group can include outlining and taking notes, receiving and following along with "quick start" lists and charts, and open conversation, discussion, and problem solving about specific applications, glitches, and bugs with the technology. These approaches can help middle-aged adults in their work productivity as long as these approaches are implemented in nonthreatening ways with peers or instructors. Consideration of prior experience, immediate goals of the user, and expeditiously cutting through superfluous information to focus on the main reasons for their technology uses are also critical considerations with middle-aged adult learners. These learners like to be recognized for the competencies they do have in a variety of areas, while being allowed to develop new competencies in ways that reduce exposure to criticism of others as much as possible.

Older Adults

Adults aged 70-75 years and older exhibit many of the same learning characteristics of the group just discussed. They also tend to be resistant to exploring tools, devices, and technologies that are new to them, as adapted from Beukelman (1990), Brookfield (1991), Schauer (1997), and others. This group of users tends to require direct personal encouragement as well as personal, patient instruction in learning new technologies and devices. Often, they may experience substantial fear of use of a new tool or device, or of doing something "wrong" with a technology that will harm it or entail costs for repairs. They may tend to use tools, devices, and technologies with which they are familiar and confident, and with which they know they have competence. They (and we all) may tend to avoid using newer devices and tools, even ones that they suspect may prove of value but that seem opaque or hard to learn to use. These items could range from push-button telephones to sewing machines to computers to microwave ovens.

Because of the rapid development of electronic, mechanical, hydraulic, computer, and other technologies over the past twenty or thirty years, many persons in this age group have been busy and occupied with other life pursuits while these technologies developed and became popular. They may have had limited time for learning and may have had little or no direct use experience with emerging complex technologies. They were in their busy family-raising time and prime work years when many of the more complex recent technologies may have developed. They also may have been put in positions where, as noted, learning of technology was threatening or aversive during these years. Now, to be comfortable with technology, they need to have one-to-one teaching and encouragement, and enough supported practice to help them increase their levels of comfort and independence with tools, devices, and machines that have not been part of their technological experience repertoire.

The preceding description of older adults' learning styles and needs with technology may be useful, but it is also too simplistic. Based on the descriptions of Brueschke (1995), Ripich (1991), and Shadden (1988), among others, the general grouping of older adults provides limited information. It should be further subdivided into three major groupings of senior adults, each of which has their own distinct characteristics. These traits can have significant impact on the learning and teaching approaches employed with men and women of these ages and should be described in more detail. Brueschke (1995) has summarized and described these distinct groupings of older adults as the Young Old (ages 65-75), the Middle Old (ages 76-85), and the Old Old (ages 86 and up). It is apparent from the outset that these arbitrary groupings of aged persons may apply primarily in industrialized countries where life expectancies are sufficient to allow the existence of groups of older persons. In some parts of the world, according to the World Health Organization, many persons do not live beyond what we in the United States consider middle age. It is difficult to make general statements about an aged population that may be minute or nonexistent. Nonetheless, some useful categorical statements can be made about persons in each of these age groups despite still limited data on the topic.

The Young Old (ages 65-75) tend to be active, physically healthy, inquisitive, and still confident in their ability to deal with the world and with change. They may be the most receptive to new tools, devices, and technologies among older learners.

In the Western world, the Middle Old (ages 76-85) are now commonly 5 to 10 years into retirement or separation from their primary employment. They are often still active and healthy, although they commonly report one or more health concerns that may interfere to some extent with their lifestyle. Hearing and vision considerations, as well as some reduction in short-term memory, motor skills, and strength, may become more evident in these years. These sensory and motoric factors can influence use and new learning of technology.

The Old Old (ages 86 and up) may now be without their spouse, may be at some distance from other surviving family members, and may not be living in their home of many years. They are probably experiencing more sensory, motoric, and memory challenges. Around 50 percent of this aged subgroup in the United States may be experiencing the onset of some type of dementia, affecting memory, cognition, personality, and behavior (Schauer, 1997). This reduction of cognitive functions can have significant impact on safe, effective use of familiar technologies. In addition, manual motor skills, including those needed for handwriting, typing, and switch or control activation, may be reduced in many members of this age group. The potential for technology use by this cohort of aged adults may seem lessened.

There is also good reason for optimism, however. A sizable and growing proportion of this age group are living independently and still have enthusiasm and capability to learn new skills with simple and complex technologies. The acquisition of new abilities with technologies of all types, ranging from socket wrenches to computers, from fax machines to blenders, is still a viable goal for many persons of this age —if they have interest in and need for learning them. Indeed, there are many documented cases of persons in the Old Old age group using and programming their phones, tablets, DVR and home televisions, or of gaining on-line skills with personal computers and enjoying the sense of community this can bring. Many aged persons have even passed the U.S. Federal Communications Commission examinations in electronic theory, governmental operating regulations and practices to become licensed amateur radio operators—sophisticated and demanding technological pursuits, indeed.

Computer usage, online games, email, social media, and other high-tech pursuits are common among older persons we have mentioned. This is borne out in the popular press and hobby magazines, as well as many emerging websites. In a more formal, scholarly manner, Lubinski and Higginbotham (1996) cover extensively the wide range of assistive communication technologies available for elderly persons. They emphasize that these technologies provide viable ways for aging persons to continue to learn, and to live a full life. The appearance of this book marks the first time that these specific areas of AT have been addressed in a comprehensive manner with a primary focus on the older user. Similarly, these authors cover many details and specifics of communication-related AT devices that affect the aged population relative to design, use, and learning. They discuss many of the essential human factors we have discussed and others particularly related to successful use of AT with older persons. The work of these authors further underscores the importance of human factors with this special (and any other) population of technology users. Both sources are valuable in providing for further details and clinical insights into successfully teaching and using technology, and specifically AT, with older persons in the age groups we have described above.

In all, our clinical practice with AT must be governed or surely at least influenced by the age of our clientele. The learning characteristics they may present toward AT for their current stage in the human life span may dramatically affect our selection of and intervention with assistive technologies of all types. In other words, simply put, another essential human factor in assistive technology is age and corresponding learning styles.

A practical, working knowledge of how people may learn and view technology at different ages allows us to control for and adapt to the characteristics they may present. It can allow us to anticipate special needs and learning traits to some extent— and to address more effectively one more potentially troublesome area of human factors in AT that could influence device acceptance and rejection.

TABLE 8-4
How Learners Across Life Span May Approach Technology

Early Childhood

Eager play and open exploration, and use varieties of control sites.

May have some fear of sounds or movement, but that is often minimal.

Little or no fear of failure or embarrassment. Tech is fun, reinforcing and engaging.

Childhood and Early Teen Years

Understand there is no, single "correct" way to do things. Flexible learners and users.

Comfortable and eager with play, learning and tech exploration. Critical, sharp users.

Young Adulthood

Skilled, critical users of more complex switches, controls, devices.

Game- and play-oriented, with hands-on learning preferred.

Some fear of failure or embarrassment. Reduced risk taking in some contexts.

Eager, enthusiastic. View technology as fun competition, socialization, challenge.

Can channel confidence with technology into increasingly work-oriented, productive applications.

Middle Adulthood

Technology viewed as means to socialize and to accomplish work. And to recreate and create

Goal and task directed. Many prefer to learn or practice in private.

Some would often rather not try than risk open failure among others.

Do not wish to be observed or closely supervised while gaining skills.

Often equate technology use with employment and performance evaluation.

Senior Adulthood

Many of same characteristics as previous group.

Fear of breaking or doing something wrong, harmful with technology. Cost concerns.

Tend to avoid and fear opaque technologies.

Can become highly skilled and enjoy new technologies with individual training and practice.

It is apparent from our limited discussion here that much more investigation into learning characteristics of persons at different ages specifically as related to AT must be undertaken. Despite some of the common difficulties in conducting research in the AT and AAC areas, such as small sample sizes and frequent reliance on normal subjects to represent special needs populations, as discussed by Bedrosian (1995), Higginbotham (1995), and Higginbotham and Bedrosian (1995), what we know already can be useful. Although these criticisms of small sample size and possible overreliance on use of larger numbers of "normal" subjects representing smaller numbers of special-needs and disabled users are valid to some extent, the related critical issue may be, more urgently, that if more standard subject-selection practices in these two areas were fully followed, virtually no research in AT and its component areas would take place. AT users are commonly a low-incidence population, and some of the major logistical problems deriving from precisely the reasons these persons need AT (mobility limitations, communication difficulties, etc.) make extended methodologies untenable. It is often difficult to gather sufficient numbers of special-needs subjects or persons who display given AT use characteristics or needs physically together in one place to conduct more traditional research methodologies. Designs involving extended information- gathering methods such as surveys by telephone, mail, e-mail, and so forth presume communication abilities (or other capabilities) that may be precisely the reasons that AT is needed in the first place by those sampled.

Although our current empirically derived knowledge of AT use and learning throughout the life span may be more limited than we would prefer, we can use what we know now to enhance what we can do now. As clinical practice and research in these areas develop a longer history and greater critical mass, we will have more data on which to make decisions, prescriptions, and predictions regarding appropriate assistive technologies across all areas of AT. Potential users of AT, with a growing wealth of technological literacy and experience, will benefit across all portions of the human life span. This is important work that must be carried out.

SUMMARY

Assistive technology can take many forms, including no-tech, low-tech, and high-tech types of AT. Whether we live with disabilities or not, we all use many forms of each to assist us in a variety of tasks each day. We can develop an experience base with devices that we have defined as technological literacy. Technological literacy relates to how a person can transfer skills learned with previous devices to new, perhaps more complex technologies. Persons with special needs, limitations, and disabilities may not be able to become as technologically literate as others in a society because of physical, sensory, or other barriers to implementation and exploration that result in limited experience with devices, tools, toys, and technologies throughout their lives. Early, planned, incremental intervention with appropriately graduated assistive technologies can help address this reduced technological literacy potential. The impact of age on learning of technology was also discussed. Throughout the human life span, we appear to have different, distinct learning characteristics and preferences. These learning characteristics at different stages in the life span should also be considered essential human factors in successful use of tech. As AT becomes better known, more widely used and ingrained in the culture, and more extensively researched, product developers, clinicians, teachers, and others will need to know more about the learning characteristics of specific individuals and populations. Correct matching of AT features with a user's specific needs and other characteristics such as gender, ethnicity, culture, and age, among other essential human factors, can be reasons that success with AT will occur. Inattention to these human factors can also be one of the reasons that AT fails. We will address this in the next chapter.

REVIEW QUESTIONS

1. Define no-, low-, and high-tech AT. Give three examples of each that you observe in your living, school or work environments.
2. What are examples of each of the above that you personally use? When and how do you use each? Specifically?

3. What advantages do lower tech methods or devices have over higher tech ones—specifically? And the reverse?
4. Is technological literacy a valid concept as described? Why or why not? Does this concept apply differently across different levels of technology? Why or why not?
5. Have you experienced ability to use a different, new, or more complex device or tool because you had experience and thus literacy with a previous, less complex device? If so, describe specifically the transfer of learning that occurred. If not, speculate on why not.
6. Observe young children interacting with devices, tools, toys, and technologies. What do you see in terms of their use, learning, enthusiasm, discouragement, and approach to further learning and use? Do the same observations with persons of other ages and compare your findings across groups on the traits listed above, plus other parameters you find relevant.
7. If you have tried to teach a new technology to someone with special needs, what impact might their age have had on the success or failure of their learning? Give specifics.
8. What age group are you currently in? Do you agree or disagree with the descriptions of learning characteristics presented here? Why? What do you personally find to be relevant technology-learning characteristics for you and your age group? Specifically?
9. What helps you learn to use new, more complex technologies? Why? What hinders your learning? Why?
10. What can you apply from your own experiences, preferences, reading, and research that may help address some of the differences in technology learning that you may see with your students, clients, or patients?
11. Describe two different research projects that should be done to further investigate technology learning across the life span. What is your rationale, hypothesis, and methodology for each?
12. Imagine that tomorrow you will teach new technology skills to a 9-year-old and a 70-year-old, in different settings.

How will your teaching approaches be similar? How might they be different? Why? Imagine and describe the same for teaching persons of different ages essential, beginning-to-intermediate skills with a new smart phone, powered wheelchair, ECU device, prosthetic limb, or any other AT device with which you may work. Describe your experiences.

9
WHEN ASSISTIVE TECHNOLOGY FAILS

When jobs are designed to match the capabilities of people, it results in better work being done and a better experience for the person doing it.
Ergoplus.com

Ergonomics is essentially fitting the workplace to the worker. The better the fit the higher the level of safety and worker efficiency.
E. Grandjean

Good kitchens are not about size. They are about ergonomics and light.
Nigel Slater

If more designers had bad backs, we would have more good chairs.
Ralph Caplan

Good ergonomics is better than cure.
Deepankar Dass

Mind your posture.
SafeNet Africa

Success with assistive technology of all types for all consumers is our consistent goal. The impact of tech on the lives of individuals can be powerful, even dramatic. Yet, AT often fails. Our efforts to make it an important, working part of our students', clients', and patients' lives can be thwarted by important variables that we may not consider or understand well— or that we may simply ignore. These variables may include human factors that are seldom mentioned, discussed, or studied at the preservice or in-service levels of professional education (Kogod, 1991; McInerney, Osher, & Kane, 1997; King, King, & King, 1997; Norman, D. 2018).

Sometimes (perhaps all too often) our best, most learned clinical efforts fail to make AT useful in consumers' lives—in perhaps one-third or more of cases (Phillips, 1991; Reed, 1997; Wirkus, 1997). Whether we call this "AT system abandonment" or "device rejection," or whether we say that some consumers and their families just give up using their AT devices and equipment because they don't like them (or "they don't work"), AT often does not accomplish the purposes which we had intended for it (Batavia, Dillars, & Phillips, 1990; Batavia & Hanuner, 1990; McInerney et al., 1997; Ourand, 1997; Phillips, 1991; Reed, 1997; Shorrock, 2017; Wirkus, 1997).

Unfortunately, AT may fail to accomplish what we hope it will for our students, clients, and patients more often than we know. The initial blush of enthusiasm that accompanies AT acquisition fades in many cases as the weeks after acquisition turn into months and years. Long-term follow up with AT often is not feasible as consumers move away and as paraprofessional or professional staff changes over time. Also, collection of long-term AT follow-up data is generally not an area of emphasis in clinical practice and research; the more immediate concerns of selection and acquisition of devices tend to receive most of our attention. AT failure is often difficult to track. When AT does fail, it often fails quietly; not much is said in many cases. It also fails at a cost to the user of lost participation in the world and in days of their lives. Further, it fails at a cost of time, effort, funds, and emotional investment for users, families, and support personnel. Although these aspects of loss are real and no doubt sizable, they are also difficult to measure validly and reliably. The variables involved in assessing and evaluating efficacy of AT intervention and use are complex and not fully delineated. When AT-related interventions fail, it is common to place the responsibility on an AT user who "gives up" too quickly, or on teachers, therapists, or families who are believed to be uncooperative, unsupportive, or inept. These may at times be valid reasons for AT failure, but these factors may also be superficial, simplistic, and too general, and may be compounded by other issues.

As is often true regarding the common consumer technologies with which we may have difficulties, we tend to blame ourselves, the device users, for the devices failing to work as intended. We tend to blame the human and to think of the user (maybe ourselves) as too inept, too clumsy, or too "dumb" to operate certain devices and machines (Norman, 1988; Sanders & McCormick, 1993; Kantowitz & Sorkin, 1983). How much more this must be the case with users who have special needs, challenges, or severe limitations. What we have covered about motivation and load factors, as summarized by Baker's basic ergonomic equation, becomes even more relevant when we are focusing on someone with technology-access differences. The insights that Fitts' law and the other human factors we have studied so far can give us regarding special access needs are invaluable. They help us to understand better some of the special challenges and daunting barriers faced by persons who have disabilities as they try to interact successfully with technology. Using tools of any type for many of these individuals is just plain hard. No wonder many of them abandon or reject the technologies they have tried and give up trying to use them further. When one also accounts for the relative inexperience of many of these persons with device use (i.e., less technological or device literacy), the likelihood for failure and for self-blame is enhanced.

When AT fails, it takes a toll on the could-have-been user. The failure or abandonment of AT says, at least in part, to the potential user and those around him that he really is as helpless as he thought. It proclaims in some ways to the user and others that she may really be as "stupid" as others have been saying for years and may not be able to use even these specially designed, carefully selected, and expensive assistive tools. The inability to use and benefit from these assistive technologies, which teachers, therapists, and family members may have selected and acquired for a user with great care, can seem to confirm further these deficiencies in AT users' lives. Learned helplessness from not being able to control the world through use of technology can turn into taught helplessness, as discussed by Roper (1997) and Norman (1988). Repeated failure experiences to use AT for its intended (and hoped-for) purposes can be powerful negative teachers, with lessons that can have lasting consequences.

So AT failure can happen and can have serious ramifications. But why does AT fail? And is it really the assistive technology itself that fails, or should we also consider other reasons? To help answer this major question, a review of current information across a variety of related fields and sources was conducted. This review included current literature in many of the fields related to AT, as well as interviews with experienced clinicians, teachers, AT consumers, and family members. Based on the information assembled and clustered via an informal "hand" factor analysis of the qualitative data assembled, three major areas of human factors related to failure of assistive technology and their component subgroupings became apparent:

1. **Factors related to the people who surround the AT user:**
    a. Evaluation and communication of diagnostic information regarding human factors
    b. Support and training
    c. Practicalities of funding
    d. Service delivery and scheduling
    e. Cultural factors

2. **Factors related to the AT user:**
    a. Matching device features to the user (not user to device)
    b. Cultural factors
    c. Age
    d. Transition among settings and devices
    e. Literacy potential
    f. Gender

3. **Factors related to the AT device itself:**

    a. Inattention to essential human factors in design and acquisition
    b. Mechanical and electrical safety and functioning
    c. Weight and size
    d. Power supply
    e. Durability and repairability

Reasons for AT failure across all component areas of the field appear to fit into one of these categories. Although these categories may not be exhaustive or precisely descriptive in all cases, the groupings presented here allow for a structuring of this complex topic. They can serve as a springboard for further discussion, research, and refinement of clinical practice. The remainder of this chapter is devoted to describing and discussing these clusters of factors in more detail, with suggestions made as possible for addressing them. Remember that these factors, and others that may not been identified or described, can all interrelate. They can—and undoubtedly do—have simultaneous impact on the success or failure, acceptance or rejection of attempted AT interventions.

Our efforts here are to organize and to make more understandable a set of interactions that likely have infinite permutations. Nonetheless, these groupings and delineating of factors give a working framework from which to proceed. As has been the intent of this entire book, a discussion of human factors related to AT failure can teach us not just to know, but also to notice— to become aware of areas we can deal with, and to be alert to factors and combinations of influences that we must still consider in pursuing efficacious AT practice.

FACTORS RELATED TO THE PEOPLE WHO SURROUND THE USER

Primary reasons that assistive technology fails center around factors related to the people who surround the technology user. Effective use of new tools of any type can be difficult for anyone. Successful integration of assistive technology into the daily lives of users with special needs, challenges, and limitations may be especially difficult. Over the long term, success with these tools for the user probably has most to do with the people who work with and care about the individual AT user. From the adequacy (or inadequacy) of initial and ongoing AT evaluations and the communication of this information to all involved, to the support and the training that helping personnel receive on a continuing basis to be able to update AT, the people who surround the AT user have a profound impact on the potential for success of AT in a person's life. Each of the subcategories listed under this major cluster of factors is described next.

Evaluation and Communication of Diagnostic Information

Perhaps the central reason that AT fails is that the user's actual needs are not truly known. The characteristics of the potential AT consumer are not explored, discovered, and considered before features of AT devices and interventions are determined. The feature-matching process, critical to success in AT, is short-circuited or ignored. This basic, important process, which helps determine and links the actual, observed needs, strengths, and characteristics of the user with the characteristics of the technology, is somehow foiled. Essential human factors are ignored. Often this oversight happens when clinicians and other professionals do not consider, and do not make the effort to find out, the true, every day, real-world needs of a client outside of the clinic room or the evaluation center. It also fails when information is not communicated or explained in meaningful ways to the family, paraprofessionals and professionals who will support the consumer's AT efforts. When the consumer is eventually expected to use the AT daily in his or her life, the AT intervention attempts fail because the AT is not acceptable, relevant, or workable in the user's real home, school, or work environments. A series of failure experiences is set in motion by inattention to detailed evaluation in environments natural for the user. This cycle may compound with inadequate communication of findings by evaluating professionals with those who will do most of the day-to-day AT work and activities with the user from the beginning (Ourand, 1997; PennTech, 1994; Reed, 1997; Wirkus, 1997).

Professionals also may tend to recommend and select for users devices or equipment with which they themselves are more familiar— rather than target the individual needs of the student, client, or patient.

This practice may especially occur if advising professionals or vendors represent a particular product, company, or point of view that they wish to promote. It can also occur when they may have a pecuniary or ego-related basis for guiding the consumer's selection of AT in a particular direction. Expensive, impressive AT may be selected that is ultimately inappropriate and irrelevant for the user. Related to this, the skill and experience levels of the evaluating professionals can affect the future success of AT. A full, thorough multidisciplinary and interdisciplinary evaluation must occur for optimal results. This does not have to be a center-based evaluation with ivory-tower experts. In fact, the field is moving more toward involvement of local teams who have expertise in AT. Whether center-based or local, these teams must be active participants in the initial evaluation of the student, client, or patient, with the most experienced members advising less experienced members and contributing substantially to the group's decisions.

AT evaluation, selection, and intervention are not processes that can be learned just from textbooks or from class notes. Effective evaluation and follow-up with AT is learned well only through training from experienced professionals. It is perfected through repeated experiences in applying and trying theoretical knowledge gained from the classroom in various settings with numerous clients—preferably in close collaboration with other, more skilled professionals. Glennen and DeCoste (1996) believe that at least 25 to 50 evaluations are required before a clinician can become minimally competent in AAC diagnostics. This is probably a realistic estimate for other component areas of AT practice as well. Professionals who have less experience and expertise must be willing to defer to the judgment and expertise of others who have more experience and must not represent themselves otherwise to consumers or families. Inexperienced, under skilled evaluators, regardless of the professions involved, generally yield inadequate evaluations. These may result in inadequate, inappropriate matching of user to technology, and may well predispose clients and attempted AT interventions to underuse, underachievement, and failure (Beukelman & Mirenda, 1992; Church & Glennen, 1992; Culp, Ambriosi, Bernings, & Mitchell, 1986; Glennen & DeCoste, 1996; Lang- ton, 1986; McMahon, 1997; Ourand, 1997; Reed, 1997; Wheatley, 2018; Wirkus, 1997).

Support and Training

All too often in AT practice, technology is selected and acquired amid a flurry of activity and enthusiasm among the professionals involved—and then is "dropped " on the user and family, who are expected to make it work. The AT fails to live up to its promise because people in the user's life simply do not share the vision for its use. They do not share ownership of making it a success—and, frankly, they often do not know what to do with it. Lack of support, communication of purposes and methods, and follow-up training for parents, families, clinicians, employers, and other supportive personnel who are important in the AT user's life can be another major contributing reason for AT failure. Either because of inattention by professionals to getting relevant data during the evaluation phase, or because of overestimating the interest, understanding, skill levels, and enthusiasm of supporting personnel, failure to offer effective clinical and technical follow-up support and training for the "background people" in the consumer's life can be another major contributor to AT failure. Overlooking the role of the people who live and work with an AT consumer—those who have daily or even moment-to-moment contact with the user—seems to be a sure way to destine the AT efforts to failure. AT users tend to have many helping, teaching, and supporting people closely involved in their lives. These people need to know how to operate, program, reprogram, modify, and even conduct minor repairs or adjustments on the assistive technology that has now been sent home with their family member, student, or client. Respect for these roles is important. Training and continuing follow-up support provided by the initial professional evaluating team and its designees can help assure that the transition of the consumer from center to home to school to work and to other AT application settings will be (and will continue to be) as successful as possible. Misunderstanding by helping personnel of the AT devices used, and simple, basic lack of information on what to expect and what to do with the device as the user grows or changes skill levels due to progression or regression in condition, can significantly influence how well a device will work for a user and for how long.

Encouraging and creating opportunities for AT users to practice device use with the more skilled and experienced professionals working with supportive people in the user's life can be a valuable way of transitioning among environments. The modeling of skills and expectations provided by experienced professionals will help others who will be taking over support of the AT to know what to expect and what to do with the devices and equipment over time. These persons can remain in contact with the original professional resources via telecommunications, occasional on-site visits, or other methods. They may then eventually become skilled and experienced enough to emerge as the local "experts" in their area—at least with certain AT items with which they have considerable experience.

Constant networking of this type, whether formal or informal, is one of the hallmarks of the AT field. Developments in the assistive technologies themselves, as well as evolving or regressing client needs, can produce rapid and dramatic changes in applications. This makes staying in constant touch with a network of experienced personnel extremely important for all of the persons involved in making AT succeed. Often, as professional staff tend to draw back from immediate, direct involvement with a case as time goes on, or as professionals move and change positions, the family of the AT user as well as his or her paraprofessionals and other helping personnel are left to continue to use and figure out the AT as the user's needs change. A network of personal and remote supportive contacts by professionals via phone, e-mail, and other methods can be extremely important in filling the information gap left by the well-intentioned but typically highly mobile cadre of professionals on whom the user, family, and others previously depended. This can be particularly pertinent with some privately practicing consultants who evaluate AT consumers and then make recommendations or even pursue acquisition of devices without being in close communication with the other important professionals, paraprofessionals, and family members in that person's life. All these people must be kept involved for best results and must be kept informed and updated in how and what to do with the assistive technology that is part of their consumer's daily life. Ideally, comprehensive academic education and clinical training of all AT-related professionals at the preservice and inservice levels can help them to be better able to multiply their impact. Their knowledge and skills can be passed on through their future training of others with whom they will collaborate for AT success across their caseload of students, clients, and patients. Device abandonment and rejection of assistive technology can be ameliorated in many cases by such proactive measures that enhance training and support to address the transitions in life that are faced by the AT user and those around him or her. These measures can help assure that the AT user has the most practical, workable technology for his or her changing needs, and that he or she and the people in the immediate environment know how to use, care for, adapt, and, if needed, pursue repairs on the devices and equipment that have been integrated in to the user's life (Brandt & Rice, 1990; Church & Glennen, 1992; Galvin & Toonstra, 1991; Glennen & DeCoste, 1996; McMahan, 1996, 1997; Ourand, 1997; PennTech, 1994; Reed, 1997; Sanford, 2016; Stainback & Stainback, 1992; Wirkus, 1997).

Practicalities of Funding

In the real world of AT use, several practical considerations can become significant factors in AT success or failure. One of the largest factors is funding. Funding is a significant enough influence in much of AT that it can drive the evaluative and intervention efforts. In short, we may often do or recommend what we believe can be funded. Professionals, consumers, and family members sometimes work from a mindset focused on "what is most affordable" rather than "what is the best" for a given consumer. Although such considerations are not supposed to influence us in the abstract, ideal world of ethical clinical practice, the truth is that cost of devices, related therapies and training, setup, configuration, repairs, and how these costs may or may not be met can have enormous impact on what devices are recommended and acquired by AT users. Devices and equipment that can be more readily afforded, and that are most likely to be covered by third-party reimbursors or other funding sources are often the devices most likely to be recommended by well-meaning professionals. They "know" they can get these items funded when others may not be, and these tend to be the items chosen for the consumer and his or her family. It also is not unusual in actual clinical practice for the type of device that requires less paperwork, less procedural struggle, and less energy investment by professional staff in acquiring funding to be the type of AT more likely to be pursued. The OT, SLP, PT, teacher, administrator, or other professional may know that a certain type of technology in a certain price range has the best chance of being approved by adjudicators who make such decisions with public and private funding sources. Therefore, even if a particular type of AT may not be the "best" for a given user, it may still be pursued for funding and acquisition if it is "close enough"—and is likely to be funded without further turmoil. Professionals involved may know from prior experiences with any given funding source that this "O.K." technology will be funded, as opposed to, perhaps, the specific, best technology that the user really needs. Items that are higher in cost or that require lengthier training and instruction periods may not have as high a probability of being funded.

The ethics codes of most professional organizations charge their members always to act in the best interests of their clients and patients. Yet the realities of funding in all clinical fields of practice operate as powerful practical forces that can cause professionals to recommend and pursue acquisition of what is adequate rather than what is best. Public school professionals can encounter a double ethical and procedural bind in this respect: They are ethically obligated to recommend the best technology and accompanying practices for their students who have special needs. But if they do so, they may obligate the school system for expenses related to AT device purchases and related therapies and may thus jeopardize their own employment and credibility with administrators in the school system. Funding of AT and its ramifications for users, families, and professionals present many difficult issues. As we move increasingly into a public-policy phase of AT practice, solutions to these types of dilemmas for consumers, families, and professionals must be found.

Currently, one partial, practical solution to this difficult area is device rental or loan. In the areas of mobility-related technologies, AAC, ECU, and others, some users' needs may be short term. In such cases, rental of equipment for a few months may be much more affordable than outright purchase. Most AT suppliers can arrange rental for AT devices and equipment for six- to eight-week periods or longer. Indeed, many public and private funding sources require a rental period of four to eight weeks for trial use and documentation of efficacy of the device before they will approve release of funds to purchase it. The Technology Assistance Act of 1988, Public Law 100-407, helped establish AT equipment and device centers throughout many states. These "tech centers" serve as repositories for demonstration and loan of specialized items to area citizens and can help provide the crucial trial run experience with devices that users, their families, and surrounding personnel must have to determine fully the appropriateness of AT before it is purchased. Use of rental and loan programs can help provide valid, extended trial use of AT and can enhance the probabilities that appropriate technologies are acquired for users who can truly benefit from them. For users with longer term or permanent AT needs, rental or loan of equipment is not a viable long-term solution. Both can still be valuable, however, in determining selection and use of the best, most appropriate AT before major efforts and funds are expended in pursuing and acquiring any make, brand, or type of devices. Documentation of successful trial use of rented or borrowed AT often is highly influential—and commonly a formal requirement—in securing funding by public, tribal and private sources for permanent acquisition of devices for consumers (Arca, 2021; Proctor and Van Zandt, 2008; WHO, 2010 & 2021).

Service Delivery and Scheduling

As the field of assistive technology has developed over the years, those with expertise in this specialized, once arcane field of practice have commonly been center-based professionals in special hospitals, rehabilitation clinics, universities, and state-operated schools or residential resource headquarters. Over the past several years, however, the availability of course work, clinical training, and research-based education in technology and AT at the undergraduate and graduate university levels has increased markedly. Likewise, the availability of in-service training and education in AT has risen for working professionals, parents, families, and paraprofessional personnel. This rising level of AT awareness, expertise, and expectations for working professionals in public schools, local hospitals, private practices, and other settings dispersed throughout urban and rural areas has allowed a more local model of AT service delivery to develop. In most areas now it is not uncommon to find local experts in AT who may have a variety of professional backgrounds and occupations such as teaching, OT, PT, SLP, special education, engineering, computer science, and other related fields. This wider availability of service providers has made AT services more widely available than in the past to many consumers. In rural and more remote areas of the United States and Canada, as well as in many other countries, services still are not at optimal levels despite recent improvements. AT consumers and their families in these circumstances still must often travel long distances to reach center-based services. This may present difficulties in training, support, and follow-up services for family and professionals in the home area but may still be the only possibility for AT services in some rural and remote areas. The other current option is that AT services simply are not available or not implemented for some consumers who might benefit from them. That, unfortunately, is still the level of service for many of those who reside in extremely rural and remote areas.

Whether AT services are delivered to child or adult consumers in rural or urban settings, scheduling and logistics of getting key people together also continue to be reasons that AT may not fulfill its promise for a given person or population. AT is a highly people-intensive enterprise; people must meet and work together in the same place and at the same time in order make support and transitioning work. The timing of AT services by skilled professionals in many service-delivery settings can also interfere with other important services in which the consumer wishes to participate. For example, in the schools it is common for the child who receives SLP, OT, or PT services to have to leave the room to participate in these therapies. These may be important, even critical times involving learning and use of AT, but it is not uncommon for these special therapies to conflict with academic instruction or other important classroom events. The child may want to stay in the room for math, reading, or a birthday party, but must be "pulled out" to receive important (but interfering) training with AAC, mobility, or other aspects of AT. Too much of this interference can reduce a child's and family's (or any other AT user's) willingness to participate in AT-related services. Such conflicts are unfortunate and should be avoided through cooperative scheduling and intercommunication, especially over the long term. These disruptions of a child's other learning experiences may have negative impact on the child's overall education if they occur at the same times over a long period. For example, if the child is always pulled out of class for AAC or other AT work during reading or math times, he or she may fall behind in learning these important curriculum areas. AT services may conflict with other important events in the lives of older children, teens, and adults, especially conflicts that interrupt a normal daily employment schedule and punctual work patterns that may have taken years to establish with a client.

Early morning or later afternoon AT services may be one partial answer to this scheduling problem, although this solution can pose scheduling difficulties for AT professionals. Perhaps the best answer overall is to use the pull- out or separate-service model for AT only when necessary, and to try to integrate services into the classroom, employment, or living setting as often as possible. In this model, the OT, PT, SLP, engineer, or other AT specialist works with the consumer in the setting where the person needs to use AT. The AT specialist attempts to teach and to give AT experience to the user in the normal setting of use that the consumer experiences, preferably involving the user's peers, teachers, support staff, family, or others who are present at that time of the day and week.

Such integration of AT services requires mature professionals who are flexible, open, and willing to be cooperative, not domineering, in their application of skills with a variety of people in a range of settings and contexts. In such cases, some professional control over all situation variables, as would be more likely in an AT clinic, is traded for more meaningful access to consumers at times and in events of their lives in which they really need to rely on AT—applied situations in which learning can be maximized. Daily or weekly contact eventually may be supplanted by periodic monitoring or consulting by the AT specialist, who can then work primarily through the teacher, job coach, or others who are part of the daily flow of life in which the consumer is immersed.

Overall, conflicts still arise often in most service-delivery settings. Much improvement and creative problem solving must occur constantly regarding scheduling. This is true whether it involves access to center-based services or whether services are provided on site at school, residential, or employment settings on a visiting or in-house basis. In a world of diverse practicalities, a myriad of related considerations may apply.

More study and investigation of best practices in this area are surely warranted. These practical, logistical aspects of learning to use AT can affect its overall success or failure with all individual consumers. Improvements and refinements continue to be necessary for optimal service delivery and maximal benefit to the consumer across all contexts of use. For further discussion of these issues, see Church and Glennen (1992), DeRuyter (1992,1995), Galvin and Toonstra (1991), Glennen and DeCoste (1996), Lloyd, Fuller, and Arvidson (1997), Ourand (1997), Rao, Goldsmith, Wilkerson, and Hildebrandt, Reed (1997), and Wirkus (1997).

Cultural Factors

Varying cultural and ethnic values, languages, belief systems, and family structures can have a profound impact on whether AT, as we in mainstream Western culture typically view it, can be successfully included in a consumer's life. An initial, vital consideration is whether the student, client, or patient with whom we are working from a high-context or a low-context culture as summarized by Platt (1996). High-context cultures value close and continued connections of family members throughout life and may often be paternal in orientation and power structure. These cultures place greater value on the surrounding "context" of the father and mother, family members, ancestors, and perceived social position of the family and extended family group as a unit and a community rather than on personal advancement or recognition of individual members. Examples of high-context cultures include many Hispanic, Asian, and Native American cultures. By contrast, low-context cultures place more value on the individual and his or her own achievement, independence, and pursuit of success and individual attainment that will increasingly set him or her apart from others in the family or society.

Low-context cultures tend to downplay the role and importance of ancestors, family members, and family status in surrounding society in general, placing responsibility and blame for success or failure on the individual's own abilities and productivity. Examples of low-context cultures include the mainstream culture in the United States, Canada, and much of northern Europe. Most of the highly educated professionals who provide AT in North America are not oriented in a high-context manner and need to be constantly alert to these basic cultural differences. We in the mainstream culture of North America are most frequently low context in our upbringing, values, orientation to the world, and view of our profession. We must recognize, however, that many of our potential AT-using students, clients, and patients will not share this background.

When an AT professional is working with persons from high-context cultures, several human factors may pertain that those of us from a low- context culture may not consider. This is probably true even if these persons are from a subculture embedded within a larger national culture, as is true with many Hispanic, Southeast Asian, and Native American families in the United States or Canada. For instance, in general, the lead decision maker and the recognized head of families from high-context cultures is the father. All decisions that may affect the children, wife, or other family members such as elderly grandparents or in-laws must first be brought to the father for his consideration, opinion, and approval. If the father is not living or is unable to act as a leader, this role is often held by the eldest surviving son in the family. AT professionals who attempt to initiate evaluations, gather or share information, conduct multidisciplinary staffing meetings, or acquire and implement assistive technology for a child or adult family member without working with and through the male head of the family are being shortsighted and unwise in their approach. Their attempts may well be rejected because they have not worked through the family hierarchy in the correct way. Even if these practices seem counter to what the modern. Western AT professional may believe, ignoring such approaches can predispose AT efforts to failure and rejection by the family. In many traditional high-context families, the father's approval must first be sought before any other actions are taken by professionals. To proceed otherwise is to be unwise, culturally blinded, and outside the expected structure that allows the consumer and his or her family to feel honored, secure, and more likely to undertake the transitions that can accompany AT use in a family by one of its members.

Additionally, independence and productivity of family members with special needs are not valued equally across all cultures. For example, in many high-context Hispanic and Southeast Asian cultures, special needs, disabilities, and limitations of a family member are seen as divine gifts or challenges that are to be met by the family. Outside interventions that draw attention to the family member or to personal achievement are not valued because they set the person apart from the rest of the family. New, expensive assistive devices or equipment may be seen as luxuries that others in the community do not have and that set a family apart from the larger context of their culture. In some cultures, the special need of a family member is viewed as "God's will," a test or even a punishment that must be borne by the family. The need for assistive technology may not be viewed as relevant because the family members themselves will be available and devoted to helping the person with special needs. They become the "biological" assistive technologies ever present to help with all of the person's needs throughout the remainder of the life span. External mechanical or electronic items are seen as expensive, de- visive, and off-putting in a family that most values unity, harmony, and group effort.

AT professionals may not agree with these beliefs and values, but ethical practice requires that these ideas be respected. Concrete suggestions for AT practice with persons from diverse cultural backgrounds include determining whom in the family to approach first regarding changes to be introduced in the life of a family member and gaining this leader's permission and trust before proceeding further with AT. Suggestions also include enlisting the assistance of a wise, respected, and trusted adult within the culturally diverse community. This venerated person may help serve as a liaison and point of entry to families of that culture that have members who can benefit from AT. In all cases of dealing with culturally diverse populations, respect for the beliefs and wishes of the family and the individual they care for must be shown, even if the AT professional finds these contrary to his or her own beliefs.

Human factors related to cultural diversity can be complex and important influences on whether our efforts with AT succeed or fail with certain persons. These differences and the bases for them are often nonintuitive. They may seem irrational and counterproductive for those of us who practice our professions from a Western, low-context mindset. Nonetheless, we must become aware of these potential differences in the populations we serve and must attempt to accommodate and work within them as much as possible while pursuing effective AT interventions. These and related issues have been further addressed by Anderson (1988), Angelo, Jones, and Kokoska (1995), Angelo, Kokoska, and Jones (1996), Atkins (1994), Conroy (1995), Duffy (1992), Fristo and Fobbs (1995), Gay (1994), Haase (1995), Hurren (1994), Lopez-DeFede (1995), Nguyen (1995), Ramirez, (2016), Spacey (2016), Smith (1995), Sotnik (1995), Stein (1995), Travelli and Vitaliti (1995), and Vue (1995), among others.

FACTORS RELATED TO THE AT USER

The second major category of reasons that AT fails centers on human factors related to the individual AT user. These factors interrelate with all of those described in the previous and the next sections. The human factors considerations in this cluster, however, relate more directly to the specific needs or characteristics of the user him- or herself. These can be critical elements in eventual AT success or failure and are the focus of this section.

Matching Device Features to User (Not User to Device)

As paraphrased from Beukelman and Mirenda (1992), the purpose of AT evaluation and trial device use is more to help discover and explore the capabilities of a potential AT user than to describe his or her deficits. Once an individual has been thoroughly evaluated by a skilled team, and factors like those described in the previous section have been considered, there are still variables concerning the user him- or herself that can affect AT success or failure. Again, the broadest view of this relationship is that expressed by Kantowitz and Sorkin (1983): "Honor thy user." In other words, in a clinical effort to determine the "best" technology and related practices for a consumer, the consumer's own needs and desires must receive foremost attention. Human factors must be considered that relate specifically to that individual user of AT. Selecting technologies and related methods, strategies, or techniques based more on the perspective of a teacher, clinician, or family member, and not on that of the AT user, is a distorted and improper way to determine what needs and characteristics of the AT user can be best accommodated by the technology.

Further guidance in this critically important process, which can be applied to specific consumers and specific devices, has been provided by the Trace Center at the University of Wisconsin-Madison. The Trace Center guidelines for universal design of products and technologies are delineated by Connell et al. (1997). They describe the concept of universal design, which they define as "the process of creating products . . . which are usable by people with the widest possible range of abilities, operating within the widest possible range of situations. ..."

Connell et al. (1997), Sanford (2016) and others have further described and summarized principles of universal design as follows:

1. **Equitable Use:** The design is useful and marketable to any group of users regardless of special needs or limitations.
2. **Flexibility in Use:** The design accommodates a wide range of individual preferences and abilities.
3. **Simple and Intuitive Use**: Use of the design is easily understood.
4. **Perceptible Information**: The design communicates necessary information effectively to the user.
5. **Tolerance for Error**: The design minimizes hazards and the adverse consequences of accidental or unintentional actions.
6. **Low Physical Effort**: The design can be used efficiently and comfortably by all users.
7. **Size and Space for Approach and Use:** Appropriate size and space, correct proxemics, is considered and provided for physical approach to and use of devices.

All these factors relate directly to our earlier foundational discussion about essential human factors such as mapping, affordances, constraints, and other fundamentals of device design as described by Kantowitz and Sorkin (1983), Norman (1988 & 2018), and Sanders and McCormick (1993), among others. With these guidelines as starting points, let us look at other considerations in AT success or failure that can surround the individual user.

Cultural Factors

As noted in the previous section, the cultural background of technology consumers, and of the people who surround them, have much to do with the success or failure of attempted AT interventions. The user must always be the primary focus— but aspects of his or her native culture, language, beliefs, and customs as they relate to the person directly, and to the family or larger community in general, must be considered. A person's own philosophy of rehabilitation, healing, and progress, as well as his or her beliefs about inclusion and participation in a larger societal role, must be honored. To force individuals to acquire and try to use technology that they do not believe in or cannot accept as part of their lives is a sure way to create AT failure. Matching even the best, most complex, and most expensive high-tech AT with users who are culturally unprepared or unwilling to accept and use such devices will still result in AT failure. The brute force of technological glitz and high cost cannot overcome deep-rooted family and cultural belief systems that may not favor AT and related activities. Effects of background culture and personal beliefs on each potential AT consumer must be considered, with AT device types, levels, and methods of instruction tailored to fit each person's needs. These must be found out. One size of AT intervention does not fit all.

---

100 words about *POCKETS IN WOMEN'S CLOTHING*

Pockets are assistive tech, making clothing useful for carrying items that otherwise occupy our hands. But full-sized, useful pockets are rare in women's jeans and

other apparel. In western culture, women began wearing pants with pockets in the 1930s. Fashionistas soon viewed pockets as unflattering, not slimming. They believed women's hands were to be kept feminine, not used for work. Women were to carry items in purses, not in pockets which could have freed up their hands. Christian Dior, 1954, stated "Men have pockets to keep things in, women [have pockets] for decoration." And that is about how things are still. Men have pants, jeans, sport coats, suitcoats and outer jackets with functional pockets for glasses, wallets, phones, combs, keys and more. Women have clutches, purses or handbags that occupy hands, with Vestigial, useless apparel pockets mostly for show. More Power to the Pocket! See https://www.theodysseyonline.com/why-are-womens-pockets-bascially-nonexistent *and* www.thefactshop.com/fashion-facts/why-are-womens-pockets-small

---

Age

The chronological age of a potential AT user can tell much about what technologies and accompanying teaching, therapy, or training strategies may match best with his or her needs. Each person is a unique individual, of course, but age in years does help predict some technology needs and feature- matching characteristics, at least on a large scale across many users. As discussed in Chapter 8, people appear to approach technology with different expectations as well as physical and mental preparedness as they pass through different portions of the life span. Attempting to match technologies and related human factors concerning switches, controls, screens, and other aspects of assistive technologies to persons of one age group that have been designed more for persons of another age group may well lead to AT failure and rejection. Even such basic human factors as switch size, selectivity, contrast, or control sensitivity can cause AT to fail if these factors are mismatched with potential AT users across age groupings. In general, it appears that design standards in the industrialized world for general consumer technologies are changing to accommodate the increasingly large portion of the population who are aging. As a larger number of AT users age, specific age-related adaptations will also become more necessary in items specifically marketed to AT consumers.

Transitions among Settings and Devices

The ability to change, adjust, and get used to AT devices is another user characteristic that can vary from person to person and from age group to age group. Predictability and structure in AT use can confer a feeling of being in charge and resultant empowerment; the opposite can turn AT use into a punishment to be avoided. Whenever new or different technology is introduced into a person's life, it causes, indeed forces, a certain amount of change to happen. This can include change in the user, change in family, school, or work routines, and changes in the physical environment of these settings. Change typically causes stress, and a basic rule of human behavior is that we try to avoid stress if possible. Stress resulting from change in AT can lead to abandonment or rejection of devices and equipment. This may especially be the result if change with AT is too frequent, too drastic, or too encompassing for a given individual. Change with AT for users of all ages (and their families and support personnel) should be foreshadowed whenever possible. Discussion, looking at pictures, contacts with other users, trial device use, and other strategies accompanied with support, training, and follow-up assistance from professionals and vendors in use of new technologies is essential. This continuing follow-up over as much time as needed for a given user can ensure that the transition is as stress-free as possible. Minimizing stress that is related to change in AT devices and equipment increases the likelihood of acceptance and meaningful use of the enabling technology over time.

Literacy Potential

The potential of an AT user for acquiring literacy skills is another major area of consideration, particularly in the AT component areas of AAC and adapted computer access. The use of appropriate technologies can help increase a user's skills as well as time on task, which in turn can help produce success for him or her with literacy. A user who shows emerging literacy abilities or who has had literacy disrupted by the onset of disability must receive AT intervention that will meet his or her specific needs for staying on task and building these skills. Blackstone and Poock (1989) have discussed literacy as a vital form of self-expression that provides access to the rest of language and communication. Determining AT users' potential for or previous skills with literacy can be critical in assuring their acceptance and continued use of the assistive communication technologies matched with them. For example, expecting AT users to rely on picture- or symbol-based vocabulary systems when they have the desire, the emerging capability, or perhaps even the fully developed ability to use a generative (spelling-based) system would be a major mistake. Allowing capable AT consumers to use AT that relies on letters, numbers, words, and other text input and output can help ensure that they have a greater opportunity to reach their full literacy and communication potential. Assistive technology can allow adults and children who have literacy capabilities to be more fully included in the communication life of their home, school, work, and the world in general. Western societies are largely text-based cultures. The ability to generate, manipulate, and work with text bestows the power to accomplish many things in a literate world—things not as fully or readily available to those who use only pictures or symbols to communicate. Writing, as we have discussed in earlier chapters, confers the ability to think on paper, and to think and refine more complex thoughts. Writing allows all of us, AT users included, to manipulate technology and, in so doing, to manipulate the world to accomplish the purposes we intend (Glennen & DeCoste, 1996). Depriving some persons of this potential through selection of inappropriate or inadequate AT devices and methods is tantamount to condemning them to separation from the rest of the literate world. Indeed, research into and further refinement of our knowledge and methods regarding how to teach literacy to AT users, especially those with severe and multiple disabilities, is an area of considerable need in the field of AT. We must realize that the capability for literacy is a valid personal characteristic, and an appropriate expectation and goal for many AT users, even those with severe communication challenges. We must find out what works best in helping them achieve literacy through

AT. This must include all aspects of literacy as described by Montgomery (1995): reading, writing, speaking, listening, and thinking. To expect emerging and competent AT users to limit achievement by withholding or not even considering giving them the tools for literacy is an open invitation for consumers to reject technology that does not meet their needs.

Gender

A final area for our discussion here that is related to users themselves is the obvious (yet often overlooked) one of gender of the user, and the AT user's legitimate preference for participation in gender-appropriate activities. From the selection of games and learning activities, to the cosmesis and affordances of AT equipment and devices, to the vocabulary and type of voice programmed onto a speech-output device, to the types of things that can be controlled by an ECU, preferences for activities, things, and behaviors that relate to expression of gender of AT consumers can affect AT acceptance or rejection. Although views of what is "appropriate" for each gender have become broader in recent years, still, children generally wish to pursue the same games and tasks that they see others of their gender pursuing. They may wish to play with toys associated with the other gender, too, but at least have the option to make the choice. To expect a reversal of this because of AT limitations, or because of our own biases about what each gender should play with or do, is to preselect for AT failure.

Teens and adults also express their gender values and roles through movement, vocabulary use, recreation, and vocational pursuits. Ultimately, successful AT must allow them to engage in the same activities and challenges as others of their gender, and with which they themselves feel comfortable. These can include being able, through the power of AT, to take a walk, play cards, fry an egg, make a phone call, send and receive e-mail, drive a car, fall in love, or care for a child. AT devices, equipment, and related methods that allow for anything less than full inclusion in the adult world will fail or at least disappoint with these consumers. At best, AT then may provide only a partial, modestly satisfying means to expression of self and gender via appearance, communication, and behavior. We have a long way to go in these areas, but at least we can be alert to them as valid user concerns and characteristics. Honoring gender-related preferences shows respect for the AT consumer, something we would expect for ourselves.

Factors related to AT consumers themselves again are many and complex. This discussion attempted to highlight some of the areas regarding AT users that are often not addressed. Support, as well as further discussion of these and other related concepts, has been provided by Beukelman (1989), Beukelman and Mirenda (1992), Church and Glennen (1992), Glennen and DeCoste (1996), Holdaway (1979), Kuenzie (1997), Spencer (1997), and others.

FACTORS RELATED TO THE AT DEVICE ITSELF

The third major category regarding why AT fails centers on human factors that surround the assistive technology itself. This cluster of factors deals with many of the essential human factors we have discussed in previous chapters as they related specifically to devices and equipment. This cluster also involves the safety, physical functioning, design, configuration, and overall usefulness of the assistive technology devices and equipment themselves. Support for this discussion is derived and synthesized from Bassock (2019), Batavia, Dillard, and Phillips (1990), Batavia and Hammer (1990), Beukelman and Mirenda (1992), Burkhart (1997), Church and Glennen (1992), Glennen and Decoste (1996), Goossens' and Crain (1992), Gullerud (1997), Jose (1983), Lynds (1997), Phillips (1991), and Yorkston, Honsinger, Dourden, & Marriner (1989), and others.

Inattention to Essential Human Factors in Design and Acquisition

These factors include device transparency-translucency-opacity, use of ordered sets of operational information, cosmesis, mappings, affordances, learned/taught helplessness, feedback, operational knowledge in the head versus in the world, constraints, and forcing functions. Each of these factors will be listed and discussed here.

- Transparency-translucency-opacity: Devices that are easier to figure out, more obvious, and more user-friendly in their

function and operation will be used more. This is as true for AT as it is for general consumer technologies.
- Use of ordered sets: Devices that have information grouped and organized in ways that we can easily sense help us to operate them more easily and more accurately. Random displays of operating controls or information make a device harder and slower to use, and tend to increase error activation.
- Cosmesis: Devices and equipment must have appearances that match well with the age, gender, and personal preferences of the user. Rejection of devices and equipment because of inappropriate cosmesis is common but avoidable.
- Mappings: The sequence and layout of controls must follow a natural, functional pattern or the device will be too confusing to operate easily. Natural mappings for individual users with challenges and special needs may not be the same as for other users. Ideally, mapping of device operation can be customized to the user to prevent rejection.
- Affordances: The design and appearance of devices can tell a lot about what they are used for, and by whom. Devices that, via their design, proclaim "handicapped" are less likely to be accepted and used by consumers sensitive to this stigma. Gender-appropriate considerations also can influence acceptance or rejection of devices for users of any age.
- Learned and taught helplessness: Devices that are too difficult to use or that are inappropriately matched to users' capabilities and needs will almost surely fail in use over time. Repeated failures with devices unsuited to individual users convey and teach the self-concept of helplessness to the user, and to those around him or her, leading to a self-perpetuating cycle.
- Feedback: Devices and equipment must offer sensory feedback to users that is meaningful in the context of their special needs and limitations. Feedback is what allows users to communicate with a device. If feedback is not present in

visual, auditory, or tactual form, the device becomes much more difficult to learn and operate as desired. Devices that do not give useful feedback are frustrating or dangerous to use and are abandoned rapidly by AT consumers and their families.

- Operational knowledge in the head versus in the world: If users must bring a complex set of knowledge and skills to the device in order to use it well, then appropriate training and practice must be provided to help them develop this operation knowledge "in the head." Many AT devices and equipment items are in this category. A better scenario, indeed, a design ideal, is to have the device itself carry most of the information needed to operate it, with this information portrayed in a transparent, readily mapped manner. Users then simply must follow the operation knowledge "in the world" that is conveyed or portrayed by the device and its surroundings.
- Constraints: Consumers with special needs and limitations may not know just what a device is for. AT must be designed so that its purpose is obvious, and so that ways that it can be used appropriately are built into the original design as much as possible. Otherwise, purposes and limitations for use must be taught expressly to the user or conveyed by the device and its surroundings as knowledge in the world.
- Forcing functions: These protective factors can be built into a device to prevent it from being used in a wrong, damaging, or unsafe manner. AT must be designed in ways that direct the user naturally to correct use, and in ways that mistakes in use can be tolerated or absorbed without significant consequences to the device, the user, or others.

One additional aspect of AT that we have not discussed earlier is the inherent two-dimensional nature of computers. This is a major limitation and perhaps a point of failure for these devices in some applications because what computers have to offer is not "real" in a three-dimensional sense. Much excitement has surrounded all of the on-screen games, activities, and adapted access to them for persons who need AT. Everything from simulations of cave exploring to frog dissections to auto repair are available, and to a certain measure, allow a type of vicarious involvement for anyone who can access and operate these programs. Although it is true that these two-dimensional on-screen experiences do offer an approximation of the actual experience in three dimensions (3-D), they are not the same thing as moving or playing in the real world. This may affect children's and adults' learning and may be a major limitation and inherent failure of on-screen solutions for some AT users' needs. Perhaps the greater availability of virtual-reality applications as well as creative uses of rehabilitation robotics can help restore more of the 3-D world to what is now literally a rather flat effort to bring the world to the computer screen for some AT consumers.

---

100 words about *ADAPTED BICYCLES AND WALKING BIKES*

Reduced mobility can mean loss of friends. Adapted bicycles and walking bikes may offer engagement, exercise, and vitality, affording fun, inclusion and fitness. New adapted bicycles and walking bikes have appealing cosmesis, conferring independent mobility while positioning users at expected levels for eye contact, conversation, and respect. Wheelchairs work well for many, but can restrict leg use, place users in submissive sitting positions, and impinge on fitness for lack of lower body exercise. Adapted and walking bikes may become appealing alternatives to wheelchairs and walkers, with attractive, sleek designs bespeaking inclusion, acceptance, competence. Online video accounts of bikes inspire with new insights & practicalities of adapted mobility. See https://www.thealinker.com **and** https://www.amerdisability.com/post/adaptive-bicycles-pave-the-way-for-riders-with-disabilites

---

Mechanical and Electrical Safety and Functioning

AT equipment of all types must ideally function safely and dependably even when it is subjected to extremes in rugged physical use, falls, dampness and moisture, and exposure to temperature extremes. Mobility devices adapted play items, AAC equipment, and other AT items that are used frequently will be subject to the jolts and bumps typical of human movement around home, and from home to school, employment, or other settings. A certain amount of physical jarring, even dropping, is unavoidable in daily life. AT items should be engineered to withstand this type of use and selected for their ability to endure rough physical handling. A broken device is useless, even dangerous, and is quickly abandoned.

In addition to the rigors of physical transport and use, AT devices must be built for exposure to moisture and water. For example, drooling, eating and drinking, use in bathroom areas or around pools, and use outdoors can expose equipment and users to the likelihood of getting wet. This can present electrical shorting hazards, potentially resulting in user harm and in damage to components of AT devices. Electrical shock and even fire hazard from shorted circuitry can also occur, particularly with devices operated from line current. These hazards also exist with battery-operated devices. Caution in design and purchase of electrically operated AT equipment is always warranted. AT equipment that will be used around water or moisture sources must be sealed and protected from water damage to prevent danger to the user and loss of function of the device. In general—although this caution may seem obvious—actual immersion of electronic AT equipment and devices in water is seldom appropriate or wise.

Exposure to extremes of heat and cold (and even blowing dust) can be significant factors for some types of AT devices—and their users. Especially in regard to AT used outside in very cold or hot climates; in climates where much change in temperature occurs; and in dusty, windy areas, the electronic, mechanical, and hydraulic components of devices must be able to survive use in such conditions without breaking or compromising reliability or safety of function. Wheelchair brakes and controls, communication device controls, adapted play items, and seating and positioning inserts may be affected, among many other possibilities. Plastic and metal parts on these and other AT devices may become too hot for safe touch if left in the sun or in a closed car. In extreme cases of exposure to heat and sun, rigid plastic overlays, plastic or rubber tubing, parts, seals, and joints can even break, splinter, warp, or melt. Potential exposure of the AT consumer's body to extremely hot or cold surfaces of such devices must be anticipated and avoided. AT users may not be able to communicate these conditions readily, so much of the responsibility for monitoring cold and heat hazards rests with the persons assisting the AT user.

Breakage of materials at extremely cold or hot temperatures can also pose a significant risk. The shattered rigid plastic cover of a switch or of an AAC overlay may result not only in interruption of function of the device, but in sharp, jagged pieces and corners that may be dangerous to the user. Such conditions of extreme cold are not uncommon in northern climates in Canada, and the United States, and other areas in which an AT consumer may be waiting outdoors for a school bus or transport van. Materials for construction of assistive devices must be chosen with extreme conditions in mind. Devices and surfaces that are jagged, broken, uncomfortable, or outright unsafe to touch will not be used well or for long.

In addition to the impact of cold or heat on AT equipment, the impact of temperature extremes on the user who may be active on a playground, at a late-fall football game, or sunbathing on the beach in the summer must be kept in mind. Hypothermia or hyperthermia, particularly among persons who have severe communication difficulties and cannot convey their discomfort when they are too hot or too cold can create potentially serious, even life-threatening situations. This can be insidiously significant for a person living with spinal cord injuries, who may not be able to sense whether his or her body is becoming too hot or too cold for safety.

Finally, as obvious as it may seem, some AT devices do not work. Whether because of design, manufacture, or configuration errors, occasionally a device or piece of equipment will not function as it is supposed to for an AT user. Careful evaluations, selection, and trial use of devices, as already discussed, can eliminate much of this concern. So can manufacturer and supplier warrantees, repair policies, and return policies—coupled with assertive advocacy skills of AT consumers, families, and professionals. An AT device that does not work correctly (or safely) should be immediately repaired or replaced. Waiting too long to do either can set in motion a cycle of "down time," leading to loss of skills and then to reduced enthusiasm for AT use and, finally, to failure of AT. This cycle can leave the user and others out of practice, and ultimately forgetting many skills they had developed with or even the primary purposes of the AT. Delays in repair or replacement of AT items that do not work must be kept minimal. Maintenance plans and prompt repair capabilities are evolving throughout the AT-related industry for equipment and devices that consumers depend on for mobility and other needs.

Weight and Size

Assistive technology must fit the user, sometimes literally. Items that are too big, too heavy, or too cumbersome will be used less. The trend in design and manufacture across all areas of AT is to more streamlined, lighter devices. This response to user needs and preferences is laudable, and will help AT to be used in contexts and settings where it could not previously go. Portability and appropriate weight are among the first considerations for users and families in choosing and accepting assistive technologies, particularly AAC devices (Glennen & DeCoste, 1996).

Power Supply

The power source for most high-tech AT is electricity. Although some golf carts and washing machines are indeed powered with gasoline engines, electricity is the usual source on which we depend for energizing most complex AT equipment and devices. Electrical current can be stored or produced in batteries that are built into a device, or it can be supplied to the device through connection to line current sources. Both types of electric power supplies have pros and cons. Battery-based power supplies offer the advantages of portability and quick replacement if needed. They usually present the disadvantages of shorter life (they have to be replaced or charged frequently) and generally lower power levels—although lower current levels can be an advantage in some toys and devices. Batteries are also disposable, which can be an advantage, but can also present significant environmental issues because of the lead, mercury, or cadmium content of some cells. Supply of electricity through line current sources is advantageous because it is continuous and can power devices that require higher current and voltage levels (desktop computers, stair lift devices, etc.). Its disadvantages include restriction of mobility (users must stay plugged in) and greater shock and fire hazard with the higher current and voltage levels.

Another disadvantage of high-tech AT and electrical power supplies of any type is that none are available in some parts of the world. Batteries or line current sources simply do not exist in all places where people may wish to use electrically operated AT. Even if electrical sources are available, they may be intermittent and undependable. This may be of considerable significance for high-current-drain AT equipment that users depend on daily and that must operate reliably and frequently. In such cases, low-tech devices and systems may be of more value than high-tech options. Indeed, low-tech backups to high-tech gear is always advisable. Some examples of these supporting measures could include a manual wheelchair to back up a powered chair or cart, or an eye-gaze AAC device made of clear plastic as a backup for an electronic speech-output device. The possibilities are numerous.

The use of wind- or solar-powered electrical sources may also be an option in some cases, although expense, practicality, and availability of these options may be daunting in many parts of the world. A truly portable, light, affordable solar power supply that could fit with and recharge batteries of laptop computers, communication devices, or other AT devices is needed now. This may be an ideal solution for electrical power needs of many developing nations, and rural and remote areas of the world. Solar (or wind) sources also may work well for persons using AT who for religious or philosophical reasons do not wish to have their homes or workplaces connected to the commercial electrical power grid. Solar power would work well with at least smaller AT equipment and devices that have lower current drain requirements and could allow "live" use or recharging of batteries on many devices. In addition to solar power, small wind-up, spring-operated electrical generators are being used with considerable success for small electronic items. This method is currently used for powering radios in remote areas of the world that are without access to batteries or line current supplies.

We trust that power supplies of these types will continue to emerge on the market and will achieve sufficient popularity to become available to and affordable by anyone. These generators could offer greater flexibility for AT applications in parts of the world where electrical power is still difficult to obtain for daily use by those people who need it. Even in the industrialized world, where the electrical supply infrastructure is highly developed, power failures still happen. Alternative power supplies, as described, may well mean that someone can talk, write, be mobile, or play even when the power grid is disrupted by storms or another catastrophe.

Durability and Repairability

All things mechanical or electrical can break. In AT, the rule of thumb may best be described this way: The more complex the device, the more prone it becomes to malfunction—and, typically, the higher the tech, the higher the complexity and cost of repair. From a practical perspective, two measures pertain.

1. Always plan on, provide for, and teach AT users at least one or more backup, no-tech, or low-tech methods for any high-tech AT devices they may acquire. Accidents, emergencies, and equipment problems do happen.
2. Select and buy equipment that can be readily repaired or replaced if needed. Check warranty policies before you purchase.

Most, though not all, major AT device manufacturers and suppliers provide for quick turnaround time and repairs of equipment, and they often provide "loaner" gear to replace the malfunctioning device. A major consideration in recommending and purchasing AT equipment should always be how well it is supported by the manufacturer and supplier, and what the practicalities of down time will be when the device does not function. These should be ascertained before selection, recommendation, and commitment of funds. Manufacturers across most component areas of AT are becoming responsive to this need and are providing repair and loaner services that seem to be improving. The examples in the AAC field cited earlier provide a target at which other manufacturers of all types of AT can aim. Consumer needs and preferences are prevailing and will continue to drive this trend.

---

100 words about *DARK PATTERNS*

Addressing human factors for hardware, software & interface devices might still not prevent all digital challenges. Harry Bugnell, expert in detecting & defining deceptive online user experiences, writes of Dark Patterns, seemingly plain language used to confuse & misdirect users in apps and websites we rely on. Types of Dark Patterns Mr. Bugnell describes are: trick questions, sneaking items into our cart or basket, roach motel irreversibility of decisions, oversharing our private information, preventing price comparisons, hidden costs, bait and switch, shaming users into compliance, disguised adds, and generating unwanted spam to friends. We must be alert. See https://www.darkpatterns.org **and** https://www.shopify.com/partners/blog/dark-patterns

---

FINAL COMMENTS: LOOKING OVER THE HORIZON

Assistive and consumer technologies are imperfect. They always will be. Is the venture still worth the effort? Yes, it surely is. AT in all its forms does work and help. It can change lives. Even the simplest of AT has the potential to help significantly if used correctly. Refinements in the human factors aspects, clinical procedures, and public policies surrounding AT are still definitely needed. Improvements are also needed in the technology itself; even the best AT is still an imperfect, inconvenient replacement for "original equipment" body systems that do not function as we would wish. The good news is that these refinements are leading to better and more widespread applications of AT. Many persons for whom technological assistance has oi\ly been a dream may increasingly find their lives enhanced through AT. Some who never knew they could move, write, talk, play, or work are finding that all these pleasures of life, and more, are within their reach now because of assistive technology. With this promise, what does the future hold for AT? Let us look over the horizon just a bit to see some of the landmarks ahead.

In many societies, being severely challenged or disabled means being poor. This has been true throughout history and across most cultures. One of the promises of AT is of improved economic viability for its users throughout their life spans. They will be able to play, learn, and to work better, and therefore eventually secure or return to meaningful, fulfilling employment. That is surely the hope for many and the reality for some now. Employment for AT consumers can confer regular use of productive time, income, and purchasing power, all of which in turn can underlie independence, dignity, and recognition as functioning human beings. Those who can participate in education and employment and who can eventually earn their own way (and pay taxes) are recognized in most industrialized societies as more fully participating citizens than those who cannot do these things. As discussed, not all persons and not all cultures share these values, but these beliefs are widespread in Western, industrialized society. These values serve as foundations for much of full participation in life in much of the world. AT can offer the tools to help more consumers attain this independence and dignity.

In our work with consumers who use AT of any type, our long-range view for each person must include how the technology he or she will acquire and learn to use can confer economic clout for the individual. In many countries, including the United States, assistive technology still is not available to all who need it. Although individuals can buy AT in the United States and in most other countries if they have the money themselves, most people do not. And that leads us back to the issue just discussed: Disabled commonly means poor, and poor means unable to pay for the AT needed to improve one's economic position. Therefore, funding of AT for persons who need it but cannot pay for it, as well as public policies regarding how AT consumers and their families can access public funds for AT when they cannot pay for it themselves, still present many deficiencies and challenges.

In Canada and many European countries, national health care plans cover the cost of AT for many users. Funding issues there are solved either by purchase of devices for individuals through government-operated care plans, or by having a variety of AT devices available for loan to and use by consumers on a state-, province-, or nationwide basis. This type of approach has many advantages in that it can connect local AT consumers, families, and professionals with the technology needed at no cost to the user. Users are given the tools they need to attain or help regain the economic power they would not have without use of AT.

Disadvantages of these types of plans, even though the AT is free to the user, may include the short supply of devices kept on hand and the lengthy waiting time involved to acquire actual use of the technologies needed. Neither the free-market approach nor the nationalized health care approach solves all the issues of funding for AT consumers. (Immediate independent wealth for all AT users is perhaps the only global answer but not a realistic one.) The future of funding for AT and ways that related policies affect real people who need AT will hold changes and refinements that may improve access to AT for everyone who needs it. This is our hope across all cultures and nations. However, the opposite could also become true. As health care costs increase, and as economies and governments worldwide rise and fall, coverage of rehabilitation-related services may be reduced in favor of funding for acute care, or in favor of balancing already tight government and industry budgets. We must be alert to this alternative future as well.

As we peer over the horizon, another landmark we see is increased universal design, as discussed by Arca (2021), Sanford (2016), and Connell et al. (1997). What we define as assistive technology and general consumer technology will further blend. As access to and design of products include more users with a broader range of needs and skills, yet still able to use the same devices, AT may become a meaningless term. As discussed in Chapter 2, all technology used by anyone can be broadly defined as assistive. Maybe we just need to broaden our view of who can be included in technology use, regardless of ability. Just as we take the design of pencils, toilet fixtures, appliances, telephones, eating utensils, and other common technologies for granted, so may consumers, designers, and manufacturers come to view technology as tools for all, with emphasis on transparency and ease of use. The evolution we have seen in home computer technology, automobiles, calculators, watches, and other consumer items suggests that this move toward inclusion of more users is making progress, still a long way to go.

Perhaps absolute universal design for all products across all humans is unrealistic, but it does present an intriguing, challenging goal for all of us. Pursuit of it will increasingly help to include many consumers who would otherwise be left out. Perhaps one day soon we will not discuss assistive technology, but rather just technology that is valuable, useful, and available for everyone based on their interests, capabilities and needs.

Finally, we live in a time of rapid change and improvement in AT. Technology has changed dramatically, and will continue to do so. Our attitudes, too, have changed about what people can and should do through consumer and assistive technologies. Opportunities and possibilities have opened wide, and on a global scale for many users.

We have seen our power to make whole, to heal, to restore, and to help many persons gain or return to fuller participation in life via AT increase immensely since around the early 1970s. The advent of the microchip and other similar technologies at that time have conferred healing, restorative tools that work. Just as in the early 1940s antibiotics emerged as powerful biochemical technologies for healing and restoring, so have electronic and mechanical technologies emerged today. These technologies have conferred much power to help on the professionals who know how to use them for their clients, and the role of these tools for consumers will expand. What we as individuals and societies come to believe technologies can and should do for persons with special needs—and that could mean anyone—is also growing.

Refinements in off-the-shelf computers and other phone, tablet, and smart devices, made by many manufacturers that will allow access for a wider range of users with a wider range of abilities are on the way. "Ubiquitous computing," wherein the microprocessor devices we rely on are built right into the clothing and accessories we wear, the cars we drive, and the rooms we inhabit will increasingly benefit consumers of AT and general technologies. The image and affordance of a computer being just a TV sitting on a typewriter will evolve, as we move inside the computer. Rehabilitation robotics, perhaps the ultimate in high-tech AT, holds much promise to help users and consumers move and manipulate our world in increasingly three-dimensional ways. Imagine the help smart robots and drones could give us all, especially if they function well for tasks we require. These and other optimistic views of what still lies over the horizon are real.

A realistic world view, however, also must include the millions of individuals who have been and continue to be injured by land mines, disabled by parasites, pandemics, illnesses and accidents, or otherwise challenged—and for whom AT does not exist. Prosthetic limbs, wheelchairs, environmental controls, or speech-output devices are not and probably will not be part of life for many persons in undeveloped countries across the planet. Persons with permanent disabilities from land mines or other sources in many countries are left economically and socially disabled for life. Marriage, family, employment, and daily survival itself are severely compromised because of their disabilities and lack of enabling technologies.

Is the promise of assistive technology for these persons as well? Some laudable efforts are already history and continuing realities. Examples are the outreach efforts of Ronald Hotchkiss at San Francisco State University to help people learn to build inexpensive wheelchairs from scrap bicycle parts for under $100. The International Society for Augmentative and Alternative Communication (ISAAC) matches developed and undeveloped Third World nations to help spread information and skills related to AAC to rural and remote areas. The Princess Diana Fund and other organized world efforts assist in detecting and destroying landmines left behind in a country. These and many other efforts are growing, helping, and adapting to newer, current needs and technologies. Low-tech, low-cost AT hacks will continue to aid, inspire and astound.

Overall, human factors and ergonomics in assistive technology for all simply come back to human beings getting the tools they need that are designed for them, and then successfully using the tools they get. True, the tools—the technologies themselves—are important and are what often initially excite and invite us into a larger, more connected and productive world.

Yet, what matters most is that humans interact successfully, safely with the technologies intended for and available to them. All else is mere acquisition of random hardware. The most essential human factor in consumer and assistive technologies is that tools of all kinds are not merely about the technology and devices themselves. Technology, low to high, simple to complex, is about people: those who need, want or are required to use it.

At their best, tools and devices we continue to design, engineer, and market must always be safe and work well when we, as the final most important step, ADD HUMAN.

REVIEW QUESTIONS

1. Has a device, tool, or technology failed when you tried to use it? Why? Was it the device itself that failed, or were other factors involved? What were they? Did you persist with the device or abandon it? Why?
2. We have described three major clusters of human factors and related variables that surround AT failure. Do you agree with these clusters and the subcategories listed? Why or why not? What would you add or subtract? Why?
3. How may the diagnostic and intervention components of AT interrelate? Can these influence technology success or failure? Why and how? For whom?
4. What practicalities of training, follow-up support, scheduling, and transitions in AT users' lives have you observed? How did these impact the use of AT? Why?
5. Have you experienced any of the practicalities mentioned here with technology in your life (dishwashers, computers, cars, answering machines, etc.)? How was your life and use of these technologies affected by these practical factors? Why?
6. Which human factors do you believe contribute the most to AT device acceptance? Why? For whom? What role should

these factors have in future design considerations for AT products? Why?
7. Which human factors contribute the most in AT device rejection or abandonment? Why? For whom? How should these be addressed in design and manufacture of other AT devices and equipment?
8. Are the factors you described in questions 6 and 7 the same for other consumer technologies? Why or why not? What impact do they have on success or failure with devices?
9. Do you agree with the definition and principles of universal design first discussed by Connell, *et al.* (1997) and others? Why or why not? What would you add or delete? Why?
10. Give an example of an inexpensive no-tech or low-tech strategy or device for each component area of AT. How could these be taught to and used by general consumers, or users of AT in developing countries? In other remote, underrecognized urban or rural areas? What could you yourself do about letting others know? Or helping adapt and design to the environment of the users?
11. Is the power to make whole, to heal, conferred by AT as great as that conferred by antibiotics and other biochemical technologies? Will this remain so? Why?
12. What future needs and trends do you see in general consumer technology and in AT? Will these be met? How, when, and by whom? Are these needs and trends really the same—or not? What role should knowledge and application of human factors play in all of these?

RESOURCES

Ideas and concepts introduced in this beginning textbook are diverse and vast. Further information related to human factors, ergonomics and assistive technology, plus links to numerous other resources and groups, are available through the organizations listed below. These resources provide bases for further exploration of human factors, ergonomics, assistive technology, and related fields. Many other valuable resources also exist in other countries and related fields.

Human Factors & Ergonomics Society (HFES)
2001 K Street, NW 3rd Floor North
Washington, DC USA 20006
Phone (202) 367-1114
FAX (202) 367-2114
Email: info@hfes.org

Secretary Europe Chapter of the HFES
Prof Dr Stephen Fairclough
Liverpool John Moores University
School of Psychology
UK
Email: secretary@hfes-europe.org

ISAAC (International Society for Augmentative and Alternative Communication)
ISAAC International Office
312 Dolomite Drive, Suite 216
Toronto, ON CANADA M3J 2N2
Tel: +1 (905) 850-6848
Fax: +1 (905) 850-6852
Email: feedback@isaac-online.org

RESNA (Rehabilitation Engineering and Assistive Technology Society of North America)
2001 K Street NW, 3rd Floor North
Washington, DC USA 2006
PHONE: (202) 367-1121
FAX: (202) 367-2121
Email: info@resna.org

American Occupational Therapy Association (AOTA)
6116 Executive Boulevard, Suite 200
North Bethesda, MD USA 20852-4929 AOTA members 1-800-SAY-AOTA (729-2682) and pressing 0. Non-members can call 301-652-6611 Email: customerservice@aota.org

Trace Research & Development Center University of Maryland College of Information Studies
Room 4105 Hornbake Building, South Wing
4130 Campus Drive
College Park, MD USA 20742 **Tel:** (301) 405-2043 **Email:** trace-info@umd.edu

United States Society for Augmentative and Alternative Communication
(USSAAC)
Email: info@ussaac.org

American Speech-Language-Hearing Association (ASHA)
ASHA National Office
2200 Research Boulevard
Rockville, MD USA 20850-3289
Local: 301-296-5700 Members: 800-498-2071 Non-Member: 800-638-8255 Fax: 301-296-8580 TTY (Text Telephone Communication Device): 301-296-5650 Email via: www.asha.org

World Health Organization (WHO) Headquarters in Geneva
Avenue Appia 20
1211 Geneva Switzerland Telephone: +41 22 791 21 11 Internet: www.who.int

BIBLIOGRAPHY & LISTING OF ONLINE SOURCES

Additional online sources are also included in the special 100 Words insets in each chapter.

Adams, J. (1982). Issues in human reliability. Human Factors, 24, 1-10.

Allen, S., Mor, V., Raveis, V., & Houts, P. (1993). Measurement of need for assistance with daily activities: Quantifying the influence of gender roles. Journal of Gerontology, 48(4), S204-S211.

Alpiner, J., & Vaughn, G. (1988). Hearing, aging, technology. International Journal of Technology and Aging, 1(2), 126-135.

Alston, R., & Mngadi, S. (1992). The interaction between disability status and the African American experience; Implications for rehabilitation counseling. Journal of Applied Rehabilitation, 23,12-15.

American Optometric Association. (1994). Using your eyes and your computer. Pamphlet. St. Louis, MO: Author.

American Psychological Association. (1996). Publication manual of the American Psychological Association (4th ed.). Washington, DC: Author.

American Psychological Association. (2014). Human Factors. Psychology Studies. Humans and Machines. Retrieved February 1, 2022. www.apa.org

Anderson, J. (1980). Cognitive psychology and its implications. San Francisco; Freeman.

Anderson, M. (2022). What is "human factors"? Retrieved February 5, 2022. www.humanfactors101.com

Anderson, P. (1988). Serving culturally diverse populations of infants and toddlers with disabilities and their families: Issues for the state. Washington, DC: National Center for Clinical Infant Programs.

Andersson, G. (1987). Biomechanical aspects of sitting; An application to VDT terminals. Behaviour and Information Technology, 6(3), 257-169.

Angelo, D., Jones, S., & Kokoska, S. (1995). Family perspective on augmentative and alternative communication: Families of young children. Augmentative and Alternative Communication, 21(9), 193-202.

Angelo, D., Kokoska, S., & Jones, S. (1996). Family perspective on augmentative and alternative communication: Families of adolescents and young adults. Augmentative and Alternative Communication, 12(1), 13-20.

Angelo, J. (1997). Assistive technology for rehabilitation therapists. Philadelphia: F. A. Davis.

Anson, D. (1990). Physical keyboard adaptations. Occupational Therapy Forum, 8(38), 1-6.

Anson, D. (1997). Alternative computer access: A guide to selection. Philadelphia: F. A. Davis.

Armstrong, J. (1978). The development of tactual maps for the visually handicapped. In G. Gordon (Ed.), Active touch. Elmsford, NY: Pergamon Press.

Amheim, D., Auxter, D., & Crowe, W. (1977). Principles and methods of adapted physical education and recreation. St. Louis: C. V. Mosby.

Arca, A. (2021). Why Study Human Factors? Transportation Research Group. University of Central Florida. Department of Psychology. Retrieved October 20, 2021. www.sciences.ucf.edu

Astrand, P., & Rodahl, K. (1986). Textbook of work physiology. New York: McGraw-Hill.

AT&T Bell Laboratories. (1983). Video display terminals. Short Hills, NJ: Author.

Atkins, B. (1994). Selected multicultural references and resources. Regional Rehabilitation Cultural Diversity Index, RSA, U.S. Department of Education.

Baddeley, A. (1986). Working memory. Oxford; Clarendon Press.

Bailey, R. (1982). Human performance engineering: A guide for systems designers. Englewood Cliffs, NJ: Prentice Hall.

Baker, B. (1986). Using images to generate speech. IEEE Biomedical Conference Proceedings, Fort Worth, TX.

Banks, W., & Boone, M. (1981). A method for quantifying control accessibility. Human Factors, 23(3), 299-303.

Barfield, W., & Robless, R. (1989). The effects of two- or three-dimensional graphics on the problem-solving performance of experienced and novice decision makers. Behaviour and Information Technology, 8, 369-385.

Barnhart, C., & Barnhart, R. (Eds.). (1989). World Book Dictionary. Chicago; World Book, Inc.

Barsley, M. (1970). Left-handed man in a right-handed world. London: Pitman.

Barth, J. (1984). Incised grids: Enhancing the readability of tangible graphs for the blind. Human Factors, 26, 61-70.

Bassock, S. (2019). Spotlight on human factors engineering: Should kids be operating medical devices independently? Retrieved October 20, 2021. www.emergobyul.com

Batavia, A., & Hammer, G. (1990). Toward a consumer-based criteria for the evaluation of assistive devices. Journal of Rehabilitation Research and Development, 27(4), 425-436.

Batavia, A., Dillard, D., & Phillips, B. (1990). How to avoid technology abandonment. Proceedings of the Fifth Annual Conference: Technology and Persons with Disabilities (pp. 43-52). Los Angeles.

Bausch and Lomb Inc. (1994). UV: The new view. Pamphlet.

Bayerl, J., Millen, D., & Lewis, S. (1988). Consistent layout of function keys and screen labels speeds user responses. Proceedings of the Human Factors Society Thirty-second Annual Meeting (pp. 344-346). Santa Monica, CA; Human Factors Society.

Bayles, K., & Kaszniak, A. (1987). Communication and cognition in normal aging and dementia. Boston: Little, Brown.

Beaton, R., & Weiman, N. (1984). Effects of touch key size and separation on menu- selection accuracy. Technical Report No. TR 500. Beaverton, OR: Tektronix Human Factors Research Laboratory.

Beck, A., & Dermis, M. (1996). Attitudes of children toward a similar-aged child who uses augmentative communication. Augmentative and Alternative Communication, 22(2), 78-87.

Bedrosian, J. (1995). Limitations in the use of nondisabled subjects in AAC research. Augmentative and Alternative Communication, 11(1), 6-10.

Beedon, L. (1992). Autonomy as a policy goal for disability and aging. Generations, 16{1), 79-81.

Belgrave, F. (1991). Psycho-social predictors of adjustment to disability in African Americans. Journal of Rehabilitation, 57, 37-40.

Bell, B. (2019). How to teach children to include classmates with disabilities. Retrieved September 21, 2021. www.enablingdevices.com

Bennett, C. (1977a). Spaces for people: Human factors in design. Englewood

Cliffs, NJ: Prentice Hall.

Bennett, C. (1977b). The demographic variables of discomfort glare.

Bennett, C. (1977c). Discomfort glare. Concentrated sources parametric study of angularly small sources. Journal of the Illuminating Engineering Society, 7(1), 2-14.

Beringer, D., & Peterson, J. (1985). Underlying behavioral parameters of the operation of touch-input devices: Biases, models, and feedback. Human Factors, 27(4), 445-458.

Bernstein, E. (1989). Liquid crystal. In World Book encyclopedia (Vol. 12, p. 347). Chicago: World Book, Inc.

Better Vision Institute. (1996). The ABCs of eyecare. Pamphlet. Washington, DC: Author.

Beukelman, D. (1989). There are some things you just can't say with your right hand. Augmentative and Alternative Communication, 5(4), 257-258.

Beukelman, D. (1990). AAC in the 1990s: A clinical perspective. In B. A. Mineo (Ed.), Augmentative and alternative communication in the next decade: Visions Conference proceedings (pp. 105-107). Wilmington: University of Delaware A. I. DuPont Institute.

Beukelman, D. (1995). Integrating augmentative communication and assistive technology into the school curriculum. Keynote address at the Seventh Annual Wisconsin Conference on AT and AAC. Eau Claire: University of Wisconsin-Eau Claire.

Beukelman, D., & Mirenda, P. (1992). Augmentative and alternative communication. Baltimore, MD: Paul H. Brookes.

Beukelman, D., & Yorkston, D. (1989). Augmentative and alternative communication application for persons with severe acquired communication disorders: An introduction. Augmentative and Alternative Communication, 5(1), 47.

Beukelman, D., Yorkston, K., Poblete, M., & Naranjo, C. (1984). Frequency of word occurrence in communication samples produced by adult communication aid users. Journal of Speech and Hearing Disorders, 49, 360-367.

Beukelman, S., Mirenda, P., Franklin, K., & Newman, K. (1992). Persons with visual and dual sensory impairments. In D. Beukelman & P. Mirenda (Eds.), Augmentative and alternative communication: Management of severe communication disorders in children and adults (pp. 291-307). Baltimore, MD: Paul H. Brookes.

Blackstone, S. (Ed.). (1986). Augmentative communication: An introduction. Rockville Pike, MD: American Speech-Language-Hearing Association.

Blackstone, S., & Poock, G. (1989). Upfront. Augmentative Communication News, 1,1-3.

Blockberger, S., Armstrong, R., O'Connor, A., & Freeman, R. (1993). Children's attitudes toward a nonspeaking child using various augmentative and alternative communication techniques. Augmentative and Alternative Communication, 9(4), 243-250.

Botten, S. (1996). Personal communication (skilled user of Liberator II and DynaVox II, Minneapolis, MN).

Boyce, P. (1981). Human factors in lighting. New York: Macmillan.

Brabyn, J., & Brabyn, L. (1982). Speech intelligibility of the talking signs. Visual Impairment and Blindness, 76, 77-7S.

Bradley, J. (1969). Optimum knob crowding. Human Factors, 11(3), 227-238.

Brandt, B., & Rice, D. B. (1990). The provision of assistive technology services in rehabilitation. Unpublished paper. Hot Springs, AR: Arkansas Research and Training Center in Vocational Rehabilitation.

Brandenburg, S. and Vanderheiden, G. (Eds.) Communication, control and computer access for disabled and elderly individuals. (Resource Book 1: Communication aids. Resource Book 2: Switches and environmental controls. Resource Book 3: Software and hardware. Waltham, MA: College Hill.

Bridgeman, B., Kirch, M., & Sperling, A. (1981). Segregation of cognitive and motor aspects of visual information using induced motion. Perception and Psychophysics, 29, 336-342.

Brookfield, S. (1991). Understanding and facilitating adult learning. San Francisco: Jossey-Bass.

Brooks, N. (1991). Users' responses to assistive devices for physical disability. Social Sciences and Medicine, 32(12), 1417-1424.

Brown, C. 1988). Human-computer interface design guidelines. Norwood, NJ: Ablex.

Brueschke, E. (Ed.). (1995). Rush-Presbyterian-St. Luke's Medical Center world medical encyclopedia. (7th ed.). Chicago: World Book, Inc.

Bullock, M. (1974). The determination of functional arm reach boundaries for operation of manual controls. Ergonomics, 17(3), 375-388.

Burke, J. (1978). Connections. Boston: Little, Brown.

Burke, J. (1985). The day the universe changed. Boston: Little, Brown.

Burkhart, L. (1997). What we have learned about augmentative communication and children functioning at young levels. Keynote address. Ninth Annual Wisconsin Conference on AAC & AT, Eau Claire.

Butterfield, N., & Arthur, M. (1996). Partners in everyday communicative exchanges. Baltimore, MD: Paul H. Brookes. (Also published by MacLennan and Petty, Pty. Ltd., Australia.)

Buzolich, M. (1988). Teaching students and their speaking peers to repair communication breakdowns. In Blackstone, S., Cassatt-James, E., & Bruskin, D. (Eds.), Augmentative communication: Implementation strategies (Vol. 5, No. 3, pp. 20-23). Rockville, MD; American Speech-Language-Hearing Association.

Campbell, M., Bush, T., & Hale, W. (1993). Medical conditions associated with driving cessation in community-dwelling, ambulatory elders. Journal of Gerontology, 48(4), S230-S234.

Cantor, M. (1985). Families: A basic source of long-term care for the elderly. Aging, 349, 8-13.

Carlson, F. (1985). Picsyms categorical dictionary. Arlington, VA: Poppin.

Carlson, F. (1986). Picsyms categorical dictionary. Unity, ME: Baggeboda Press.

Carlson, F. (1992). DynaSyms™: Symbol vocabulary system of DynaVox and DynaVox II. Philadelphia, PA: Sentient Systems Technologies, Inc.

Carter, R. (1978). Knobology underwater. Human Factors, 20, 641-647.

Cerella, J. (1985). Information processing rates in the elderly. Psychological Bulletin, 98, 67-83.

Chaffin, D., & Andersson, G. (1984). Occupational biomechanics. New York; Wiley.

Chapanis, A. (1967). The relevance of laboratory studies to practical situations. Ergonomics, 10(5), 557-577.

Chapanis, A. (1976). Engineering psychology. In M. D. Dunnette (Ed.), Handbook of industrial and organizational psychology. Chicago: Rand McNally.

Chapanis, A. (1991). To communicate the human factors message, you have to know what the message is and how to communicate it. Bulletin of the Human Factors Society, 34, 1-4.

Charlebois-Marois, C. (1985). Everybody's technology. Montreal: Charlecoms.

Chomsky, N. (1968). Language and mind. New York: Harcourt Brace Jovanovich.

Christ, R. (1975). Review and analysis of color-coding research for visual displays. Human Factors, 17, 542-570.

Church, G., & Glennen, S. (1992). The handbook of assistive technology. San Diego, CA: Singular Publishing Group.

Citing datafiles and Internet sources. (1997). Eau Claire: W. D. McIntyre Library, University of Wisconsin-Eau Claire.

Clark, M., & Gaide, M. (1986). Choosing the right device. Generations, 20(3), 18-21.

Clark, S., & Kelley, S. (1992). Traditional Native American values: Conflict or concordance in rehabilitation? Journal of Rehabilitation, April-June, pp. 23-28.

Clay, M. (1985). The early detection of reading difficulties (rev. ed.). Portsmouth, NH; Heinemann.

Cline, D., Hofstetter, H., & Griffin, J. (1980). Dictionary of visual science. Radnor, PA: Chilton Book Company.

Connell, B., Jones, M., Mace, R., Mueller, J., Mullick, A., Ostroff, E., Sanford, J., Steinfeld, E., Story, M., & Vanderheiden, G. (1997). Universal design: The Trace Center definition and seven principles of universal design. Trace Center Overview. Madison: UW-Madison.

Conrad, R., & Longman, D. (1965). A standard typewriter versus chord keyboard—An experimental comparison. Ergonomics, 8, 77-88.

Conroy, J. (1995). Outreach to Native American populations requires an awareness of human diversity and cultural sensitivity. Ln L. Travelli Vitaliti & E. Bourland (Eds.), Project Reaching Out: Proceedings of the Forum on Human Diversity (pp. 140-167). Arlington, VA: RESNA.

Cook, A., & Hussey, S. (1995). Assistive technologies: Principles and practices. St. Louis, MO: Mosby.

Corcoran, M. (2018) When Alexa Can't Understand You. SLATE. Retrieved October 11, 2021. www.slate.com

Coward, R., Bull, C., Kukulka, G., & Galliher, J. (Eds.) (1994). Health services for rural elders. New York: Springer.

Coward, R., & Lee, G. (Eds.). (c 1993). The elderly in rural society: Every fourth elder. New York: Springer.

Crossman, E., & Goodeve, P. (1983). Feedback control of hand movements and Fitts' law. Quarterly Journal of Experimental Psychology, 35A, 251-278.

Culp, D., Ambiosi, D., Beminger, T., & Mitchell, J. (1986). Augmentative communication aid use: A follow-up study. Augmentative and Alternative Communication, 2, 19-24.

Cunningham, J. (1980). Cable television (2nd ed.). Indianapolis, IN: Howard W. Sams.

Czaja, S. (1988). Microcomputers and the elderly. In M. Helander (Ed.), Handbook of human-computer interaction. Amsterdam: North-Holland.

Davila, R. (1991). Goals for improved services to minority individuals with disabilities. OSERS News in Print, 3(4), 2-5. Washington, DC: U.S. Department of Education.

Denning, J. (1996). Personal communication (DynaVox consultant and parent of AT user, Minneapolis, MN).

DeRuyter, F. (1992). The importance of cost benefit analysis in AAC. In Consensus validation conference: Resource papers. Washington, DC: National Institute on Disability and Rehabilitation Research.

DeRuyter, F. (1995). Evaluating outcomes in assistive technology: Do we understand the commitment? Assistive Technology, 7, 3-8.

Design in action. Success. (c 1995). 39(8), 16.

Dhillon, B. (1986). Human reliability with human factors. New York: Pergamon Press.

Diamond, J. (1997). The curse of QWERTY. Discover, April, 34-42.

DiChristina, M. (1997). Liquid crystal displays. Popular Science, March, 86.

Dionne, E. (1984). Carpal tunnel syndrome; Part I—The problem. National Safety News, 42-45.

Dorland's illustrated medical dictionary. (1985). Philadelphia; W. B. Saunders. (John P. Friel, dictionary editor for publication).

Downing, & Sanders, M. (1987). The effects of panel arrangement and locus of attention on performance. Human Factors, 29(5), 551-562.

Drury, C. (1975). Application of Fitts' law to foot-pedal design. Human Factors, 17, 368-373.

Ducharme, R. (1977). Women workers rate "male" tools inadequate. Human Factors Society Bulletin, 20(4), 1-2.

Duffy, J. (1992). ESL and alienation: The Hmong of St. Orien. Wausau Hmong Association.

Duffy, S. (1994). Rural America: Wide open spaces where the disability issues resemble those in urban areas. Mainstream, June-July 17-21.

Dvorak, A. (1943). There is a better typewriter keyboard. National Business Education Quarterly, 12, 51-58.

Early intervention services birth to 3; Because all kids are special. (Pamphlet). (1997). Cumberland, WI; Northern Pines community programs.

Easterby, R. (1970). The perception of symbols for machine displays. Ergonomics, 13,149-158.

Eastman Kodak Company. (1986). Ergonomic design for people at work (Vol. 2). New York; Van Nostrand Reinhold.

Elder, P., and Goosens', C. (1994). Engineering daily living and vocational training environments for interactive symbolic communication. Birmingham, AL: Southeast Augmentative Communication Publications.

Ellis, J., & Dewar, R. (1979). Rapid comprehension of verbal and symbolic traffic sign messages. Human Factors, 21,161-168.

Ennes, H. (1971). Television broadcasting equipment, systems, and operating fundamentals. Indianapolis: Howard W. Sams.

Farewell to the clunky look. (1988). U.S. News & World Report, 104(23), 56.

Federal aviation Administration. (2016). Avoid the Dirty Dozen: 12 Common Causes of Human Factors Errors. Retrieved October 4, 2021. www.faasafety.gov

Fiedler, C., & Simpson, R. (1987). Modifying the attitudes of nonhandicapped high school students toward handicapped peers. Exceptional Children, 53, 342-349.

Fisher, D., & Tan, K. (1989). Visual displays: The highlighting paradox. Human Factors, 31,17-30.

Fitts, P. (1954). The information capacity of the human motor system in controlling the amplitude of movement. Journal of Experimental Psychology, 47, 381-391.

Fitts, P., & Peterson, J. (1964). Information capacity of discrete motor responses. Journal of Experimental Psychology, 67, 103-112.

Flax, M., Golembiewski, D., & McCauley, B. (1993). Coping with low vision. San Diego, CA; Singular Publishing Group.

Flippo, K., Inge, K., & Barcus, M. (Eds.). (1995). Assistive technology. Baltimore, MD: Paul H. Brookes.

Fortner, K. (1997). Personal communication (assistive technology consultant and special educator, Minneapolis, MN).

Forbes, W., Sturgeon, D., Hayward, L., Awwani, N., & Dobbins, P. (1992). Hearing impairment in the elderly and the use of assistive hearing devices; Prevalences, associations, and evaluations. International Journal of Technology and Aging, 5(1), 39-61.

Ford, A., & Roy, A. (1991). Impaired and disabled elderly in the community. American journal of Public Health, 82(9), 1207-1209.

Friel, D. (1997). Personal communication (assistive technology specialist. Courage Center, Golden Valley, MN).

Fristo, M., & Fobbs, J. (1995). The bridge across the divide; Delivering assistive technology and rehabilitation service to culturally diverse consumers. In L. Travelli Vitaliti & E. Bourland (Eds.), Project Reaching Out: Proceedings of the Forum on Human Diversity (pp. 182-198). Arlington, VA; RESNA.

Fuller, D., & Lloyd, L. (1987). A study of physical & semantic characteristics of a graphic symbol system as predictors of perceived complexity. Augmentative & Alternative Communication, 3, 26-35.

Fuller, D., & Lloyd, L. (1992). Effects of configuration on the paired-associate learning of Bliss symbols by preschool children with normal cognitive abilities. JSHR, 35, 1376-1383.

Galvin, J., & Toonstra, M. (1991). Adjusting technology to meet your needs. Proceedings of the Sixth Annual Conference on Technology and Persons with Disabilities (pp. 267-273). Los Angeles, CA.

Garcia, S., & Malkin, D. (1993). Toward defining programs and services for culturally and linguistically diverse learners in special education. Teaching Exceptional Children, 52(26), 1.

Gay, J. (1994). Human diversity: Reaching the elderly population. Paper presented at the Forum on Human Diversity, Washington, DC.

Gay, J., (1995). Human diversity: Reaching the elderly population. In L.

Travelli Vitaliti and E. Bourland (Eds.) Project Reaching Out: Proceedings of the Forum on Human Diversity. (pp. 59-96) Arlington, VA: RESNA.

Geldard, F. (1957). Adventures in tactile literacy. American Psychologist, 12, 115-124.

Gelfand, D. (1994). Aging and ethnicity: Knowledge and services. New York: Spring.

Gian, K., & Hoffman, E. (1988). Geometrical conditions for ballistic and visually controlled movements. Ergonomics, 31(5), 829-839.

Ginsburg, A., Evans, D., Carmon, M., Owsley, C., &c Mulvanny, P. (1984). Large-scale norms for contrast sensitivity. American journal of Optometry and Physiological Optics, 61, 80-84.

Gitlin, L., & Levine, R. (1992). Prescribing adaptive devices to the elderly: Principles for treatment in the home. International Journal of Technology and Aging, 5(1), 107-120.

Glennen, S., & Decoste, D. (1996). The handbook of augmentative and alternative communication. San Diego, CA; Singular Publishing group.

Goossens', C. (1989). Aided communication intervention before assessment: A case study of a child with cerebral palsy. Augmentative and Alternative Communication, 5,14-26.

Goossens', C., Berg, C., Lane, S., & Crain, S. (1987). Guidelines for establishing a manual point. Handout from sectional presentation at Closing the Gap International Conference, Minneapolis, MN.

Goosens', C. & Crain, S. (1992). Utilizing switch interfaces with children who are severely physically challenged. Austin, TX: Pro-Ed.

Goossens', C., Crain, S., & Elder, P. (1992). Engineering the preschool environment for interactive symbolic communication. Birmingham, AL: Southeast Augmentative Communication Publications.

Gorenflo, C., & Gorenflo, D. (1991). The effects of information and augmentative communication technique on attitudes toward nonspeaking individuals. journal of Speech and Hearing Research, 34,19-26.

Gould, J., & Grischkowsky, N. (1984). Doing the same work with hard copy and with cathode-ray tube (CRT) computer terminals. Human Factors, 26, 323-337.

Grabowski, D. (1997). Personal communication (electronics technician at University of Wisconsin-Eau Claire).

Grandjean, E. (1987). Ergonomics in computerized offices. London; Taylor & Francis.

Grandjean, E. (1988). Fitting the task to the man. London; Taylor & Francis.

Grandjean, E., Hunting, W., & Piderman, M. (1983). VDT workstation design; Preferred settings and their effects. Human Factors, 25,161-175.

Grant, C., & Brophy, M. (1994). An ergonomics guide to VDT workstations. Fairfax, VA: American Industrial Hygiene Association.

Gray, D., Quatrano, L., & Lieberman, M. (1997). Designing and using assistive technology: The human perspective. Baltimore, MD: Paul H. Brookes.

Green, P. (1984). Driver understanding of fuel and engine gauges. Technical Paper Series 840314. Warrendale, PA; Society of Automotive Engineers.

Grieco, A. (1986). Sitting posture. An old problem and a new one. Ergonomics, 29(3), 343-362.

Grillner, S. (1985). Neurobiological bases of rhythmic motor acts in vertebrates. Science, 228,143-149.

Guillen, M. (1995). Five equations that changed the world. New York: Hyperion.

Gullerud, R. (1997). Personal communication. Administrator and assistive technology user. Eau Claire, WI.

Haase, S. (1995). Disability and technology-related assistance. Responsive approaches to the rehabilitation of Native Americans: Reactor statement. In L. Travelli Vitaliti & E. Bourland (Eds.), Project Reaching Out: Proceedings of the Forum on Human Diversity (pp. 168-181). Arlington, VA: RESNA.

Harpster, J., Freivalds, A., Shulman, G., & Leibowitz, H. (1989). Visual performance on CRT screens and hard copy displays. Human Factors, 31, 247-257.

Hawking, S. (1990). Personal communication. Golden Valley, MN. Courage Center Award recipient and guest speaker.

Hayflick, L. (1996). How and vhy we age. New York: Ballantine Books.

Hays, W. (1988). Statistics (4th ed.). New York: Holt, Rinehart & Winston.

Healthy People 2000. (1990). DHS Publication No. (PHS) 91-50213, Superintendent of Documents. Washington, DC: U.S. Government Printing Office.

Heinlein, R. (1958). Have Spacesuit Will Travel. New York: Charles Scribner's Sons.

Helander, M. (2006) A Guide to Human Factors and Ergonomics. 2nd ed. Boca Raton, FL: Taylor & Francis.

Helander, M., & Rupp, B. (1984). An overview of standards and guidelines for visual display terminals. Applied Ergonomics, 15(3), 185-195.

Helge, D. (1984). Models for serving rural students with low incidence handicapping conditions. Exceptional Children, 50(4), 313-324.

Henke, C. (1998). No longer just rehab aids. RX Home Care: The journal of Home Health Care and Rehabilitation, 10 (11), 34-36.

Herbenson, J., & Sather, T. (1996). AAC Access method cosmesis as determined by fifth grade males. Poster presentation of empirical research at Fourth Annual UW-Eau Claire Student/Faculty Collaborative Research Day, Eau Claire, WI.

Hetzroni, O., & Harris, O. (1996). Cultural aspects in the development of AAC users. Augmentative and Alternative Communication, 12(1), 52-58.

Higginbotham, D. J. (1995). Use of nondisabled subjects in AAC research: Confessions of a research infidel. Augmentative and Alternative Communication, 12(1), 2-5.

Higginbotham, D. J., & Bedrosian, J. (1995). Subject selection in AAC research: Decision points. Augmentative and Alternative Communication, 2 2(1), 11-13.

Holdaway, D. (1979). The foundations of literacy. New York: Ashton, Scholastic.

Holland, B., & Falvo, D. (1990). Forgotten: Elderly persons with disability— A consequence of policy. Journal of Rehabilitation, 56(2), 32-35.

Holm, M., & Rogers, J. (1991). High, low, or non-assistive technology devices for older adults undergoing rehabilitation. International journal of Technology and Aging, 4(2), 153-162.

Holvey, D. (1972). The Merck manual of diagnosis and therapy (12th ed.). Rahway, NJ: Merck, Sharp & Dohme Research Laboratories.

Denno, S., et al. (1992). Human factors design guidelines for the elderly and people with disabilities. Honeywell. Retrieved February 2, 2022. www.cmu.edu

Howell, S. (1980). Designing for aging: Patterns of use. Cambridge, MA: MIT Press.

Howett, G. (1983). Size of letters required for visibility as a function of viewing distance and viewer acuity. NBS Technical Note 1180. Washington, DC: National Bureau of Standards.

Hull, R., & Griffin, K. (1989). Communication disorders in aging. Beverly Hills, CA: Sage Publications.

Human Factors and Ergonomics Society directory and yearbook. (1997-1998). Santa Monica, CA: Human Factors and Ergonomics Society.

Human Factors Society. (1988). American national standard for human factors engineering of visual display terminal workstations. ANSI/HFS 100-1998. Santa Monica, CA: Human Factors Society.

Hurren, D. (1995). Reaching the elderly; Reactor statement. In L. Travelli Vitaliti & E. Bourland (Eds.), Project Reaching Out: Proceedings of the Forum on Human Diversity (pp. 97-104). Arlington, VA: RESNA.

Iavecchia, J., Iavecchia, H., & Roscoe, S. (1988). Eye accommodation to head-up virtual images. Human Factors, 30(6), 689-702.

Ilg, R. (1987). Ergonomic keyboard design. Behaviour and Information Technology. 6(3), 303-309.

Indianhead Federated Library System. (1996). Your changing vision. Eau Claire, WI: Author.

Institute of Medicine. (1991). Disability in America: Toward a national agenda for prevention. Washington, DC: National Academy Press.

Introduction to Human Factors Engineering. (2018). Part of engineering series. YouTube video. Retrieved November 9, 2021. www.youtube/watch?v=8JnflkwaDoc

Jagacinski, R., Flach, J., & Gilson, R. (1983). A comparison of visual and kinesthetic-tactual displays for compensatory tracking. IEEE Transactions on Systems, Man & Cybernetics, 13,1103-1112.

James, G., & Armstrong, J. (1975). An evaluation of a shopping center map for the visually handicapped, journal of Occupational Psychology, 48,125-128.

James, P., & Thorpe, N. (1994). Ancient inventions. New York: Ballantine Books.

Jensen, C., Schultz, G., & Bangerter, B. (1983). Applied kinesiology and biomechanics. New York: McGraw-Hill.

Johnson, C. (1995). Personal communication (audiologist and visiting minority scholar. University of Wisconsin-Eau Claire).

Johnson, J., Baumgart, D. Helmstetter, E., & Curry, C. (1996). Augmenting basic communication in natural contexts. Baltimore, MD: Paul H. Brookes.

Jose, R. (1983). Understanding low vision. New York: American Foundation for the Blind.

Joyce, M. (1993). Greater freedom for the elderly. Design News, 49(11), 124-127.

Kahneman, D. (1973). Attention and effort. Englewood Cliffs, NJ: Prentice-Hall.

Kantowitz, B., & Sorkin, R. (1983). Human factors: Understanding people-system relationships. New York: Wiley.

Karp, J. (1987). Personal communication. Project Active consultant, Elmwood, WI.

Karpman, R. (1992). Problems and pitfalls with assistive devices. Topics in Geriatric Rehabilitation, 8(2), 1-5.

Kart, C., Metress, E., & Metress, S. (1988). Aging, health, and society. Boston: Jones and Bartlett.

Katz, S., Ford, A., Moskowitz, R., Jackson, B., & Jaffe, M. (1963). Studies of illness in the aged: The Index of ADL—A standardized measure of biological and psychosocial function, journal of the American Medical Association, 185, 914-919.

Kelso, J. (1982). Human motor behavior: An introduction. Hillsdale, NJ: Lawrence Erlbaum Associates.

Kenney, R. (1982). Physiology of aging: A synopsis. Chicago: Yearbook Medical Publishers.

King, D. (2022). Gravedigger's Daughter – growing up rural. Hartland, WI: WWA Press.

King, K. A. (1999). Personal communication with group home residents and staff. Hudson, WI.

King, S., Rosenbaum, P., Armstrong, R., & Milner, R. (1989). An epidemiological study of children's attitudes toward disability. Developmental Medicine and Child Neurology, 31, 237-249.

King, T. (1990). An improvised eye-pointing communication system for temporary use. Language, Speech & Hearing Services in the Schools, 21(2), 116-117.

King, T. (1991). A signaling device for non-oral communicators. Language,

Speech & Hearing Services in the Schools, 22(1), 116-117.

King, T. (1997). Getting started in AT and AAC. Sectional presentation at the Ninth Annual Wisconsin Conference on AT and AAC, Eau Claire.

King, T., King, A., & King, S. (1997). Essential human factors in assistive technology. A sectional presentation at the Fourth Annual Wisconsin Technology Access Conference, Milwaukee.

King, T., Noel, E., Rounds, L., & Bergo, T. (1991). Enhancing mouthstick use: Making a mouthstick—and making the most of it. Closing the Gap, 11, 22-23.

King, T., Schomisch, L., «& King, A. (1996). Essential human factors in assistive technology. Sectional presentation at the Eighth Annual Wisconsin Conference on Assistive Technology and Augmentative and Alternative Communication, Eau Claire.

King, T., Schomisch, L., & King, A. (1996). Essential human factors in assistive technology. Sectional presentation at 13th Annual Closing the Gap International Conference, Minneapolis.

King, V. H. (1998). Personal communication with nursing home resident. Eau Claire, WI.

King, V. (1997). Personal communication (professor of optometry, Ferris State University, Big Rapids, MI).

Knave, B. (1984). Ergonomics and lighting. Applied Ergonomics, 25(1), 15-20.

Kochhar, D., & Barash, D. (1987). Status displays in automated assembly. Applied Ergonomics, 28(2), 115-124.

Kogod, S. (1991). A workshop for managing diversity and disability: A workshop manual. San Diego, CA: Pfeiffer.

Kohl, G. (1983). Effects of shape and size on knobs on maximal hand-turning forces applied by females. The Bell System Technical Journal, 62,1705-1712.

Koppenhaver, D., Evans, D., & Yoder, D. (1991). Childhood reading and writing experience of literate adults with severe speech and motor impairments. Augmentative and Alternative Communication, 7, 20-33.

Kroemer, K. (1970). Human strength: Terminology, measurement, and interpretation of data. Human Factors, 22(3), 297-313.

Kryter, K. (1983). Presbycusis, sociocusis and nosocusis. Journal of the Acoustical Society of America, 73(6), 1897-1917.

Kryter, K. (1985). The effects of noise on man (2nd ed.). Orlando, FL: Academic Press.

Kuenzie, P. (1997). Personal communication (social worker, St. Croix County, WI).

LaBuda, D. (1989). Aging and design: The issues and background. SOMA Engineering for the Human Body, 3(1), 7-13.

Langton, T. (1986). Personal communication (director. Center for Rehabilitation Technology, University of Wisconsin-Stout, Menomonie, WI).

Laveson, J., & Meyer, R. (1976). Left out "lefties" in design. Proceedings of the Human Factors Society Twentieth Annual Meeting (pp. 122-125). Santa Monica, CA: Human Factors Society.

Leal-Idrogo, A. (1993). Vocational rehabilitation of people of Hispanic origin. Journal of Vocational Rehabilitation, 3, 27-37.

Leland, M., & Schneider, M. (1982). Rural rehabilitation: A state of the art. Washington, DC: National Institute for the Handicapped.

Lendle, D., & Maro, V. (1996). Personal communication (enabling technology. Northern Wisconsin Center for the Developmentally Disabled, Chippewa Falls, WI).

Leonard, S., & Romanowski, J. (1995). Morse code: Another way to write. Handout and title for presentation at seventh annual Wisconsin Conference on Assistive Technology and Augmentative and Alternative Communication, Eau Claire, WI.

Leung, P. (1993). A changing demography & its challenge. Journal of Vocational Rehabilitation, 3, 3-11.

Light, J. (1989). Toward a definition of communicative competence for individuals using augmentative and alternative systems. Augmentative and Alternative Communication, 5, 137-144.

Light, J., & Binger, C. (1997). Building communicative competence with individuals who use augmentative and alternative communication. Baltimore, MD: Paul H. Brookes.

Lloyd, L., Fuller, D., & Arvidson, H. (1997). Augmentative and alternative communication: A handbook of principles and practices. Boston: Allyn and Bacon.

Lopez-DeFede, A. (1995). Outreach and service provision to persons with disabilities residing in rural and sparsely populated areas. In L. Travelli Vitaliti & E. Bourland (Eds.), Project Reaching Out: Proceedings of the Forum on Human Diversity (pp. 105-128). Arlington, VA: RESNA.

Lorenz, D. (1997). Personal communication. Optometrist, Eau Claire, WI.

Lubinski, R., & Higginbotham, J. (1996). Communication technologies for the elderly. San Diego, CA: Singular Publishing Group.

Lueder, R. (1991). Seating a new worker. In J. Sweere (Ed.), Chiropractic family practice. Rockville, MD: Aspen.

Lunn, R., & Banks, W. (1986). Visual fatigue and spatial frequency adaptation to video displays of text. Human Factors, 28, 457-464.

Lunzer, F. (1989). Small gadgets that can change lives. U.S. News & World Report, 106(9), 58-60.

Lynch, E., & Hanson, M. (1992). Developing cross-cultural competence: A guide for working with young children and their families. Baltimore, MD: Paul H. Brookes.

Lynds, J. (1997). Personal communication (engineer and product developer/manufacturer, Minneapolis, MN).

Mandler, J. (1984). Stories, scripts, and scenes: Aspects of schema theory. Mahwah, NJ: Laurence Erlbaum Associates.

Mann, W. (1989). Consumer role in statewide planning for access to assistive technology. Journal of Rehabilitation and Development, 26(Suppl.), 463.

Mann, W., Hurren, D., Karuza, J., & Bentley, D. (1993). Needs of home-based older visually impaired persons for assistive devices. Journal of Visual Impairment and Blindness, 87(3), 106-110.

Mann, W., Hurren, D., & Tomita, M. (1993). Comparison of assistive device use and needs of home-based older persons with different impairments. The American Journal of Occupational Therapy, 47(11), 980-987.

Mann, W., Hurren, D., & Tomita, M. (1994). Assistive device needs of home-based elderly persons with hearing impairments. Technology and Disability, 3(1), 47-61.

Mann, W., Karuza, J., Hurren, D., & Tomita, M. (1992). Assistive devices for home-based elderly persons with cognitive impairments. Topics in Geriatric Rehabilitation, 8(2), 35-52

Mann, W., Karuza, J., Hurren, D., & Tomita, M. (1993). Needs of home-based older persons for assistive devices. Technology and Disability, 2(1), 1-11.

Mann, W., Tomita, M., Packard, S., Hurren, D., & Creswell, C. (1994). The need for information on assistive devices by older people. Assistive Technology, 6(2).

Maro, J. (1993). Personal communication (speech-language pathologist and AAC consultant, Eau Claire, WI).

Marotz-Baden, R., Hennon, C., & Brubaker, T. (Eds.). (1988). Families in rural America: Stress, adaptation and revitalization. St. Paul, MN; National Council on Family Relations.

Marshall, C., Johnson, S., & Lonetree, G. (1993). Acknowledging our diversity, vocational rehabilitation and American Indians. Journal of Vocational Rehabilitation, 3, 12-19.

Martin, W., Frank, L., Minkler, S., & Johnson, M. (1988). A survey of vocational rehabilitation counselors who work with American Indians. Journal of Rehabilitation Counseling, 29(4), 29-34.

Marvin, C., Beukelman, D., Brockhaus, J., & Kast, L. (1994). What are you talking about? Semantic analysis of preschool children's conversational topics in home and preschool settings. Augmentative and Alternative Communication, 10, 75-86.

Marwick, C. (1993). Older people now more able-bodied than before. JAMA, 260(18), 2333-2337.

Masoro, E. (Ed.). (1995). Handbook of physiology, section 11: Aging. New York: Oxford University Press.

Matthews, M. (1987). The influence of colour on CRT reading performance and subjective comfort under operational conditions. Applied Ergonomics, 18, 323-328.

Matthews, M., Lovasik, & Mertins, K. (1989). Visual performance and subjective discomfort in prolonged viewing of chromatic displays. Human Factors, 31, 259-271.

Maxim, J., & Bryan, K. (1994). Language of the elderly: A clinical perspective. San Diego: Singular Publishing Group.

McEwen, I., & Lloyd, L. (1990). Some considerations about the motor requirements for manual signs. Augmentative and Alternative Communication, 6, 207-216.

McInerney, M., Osher, D., & Kane, M. (1997). Improving the availability and use of technology for children with disabilities: Final report. Washington, DC: Chesapeake Institute of the American Institutes for Research.

McMahan, D. (1996). Research reveals the power of Morse Code in rehab settings for MORSELS. Newsletter of the Morse 2000 Worldwide Outreach, UW-Eau Claire, 2(2), 1-2.

McMahon, D. (1997). Personal communication (speech-language pathologist and assistive technology specialist, Lexington, KY).

Mecham, M. (1996). Cerebral palsy. Austin, TX: Pro-Ed.

Meeting the unique needs of minorities with disabilities. (1993). National Council on Disability, Report to the President and the Congress. April.

Meister, D. (1989). Conceptual aspects of human factors. Baltimore, MD: Johns Hopkins University Press.

Melnick, W. (1979). Hearing loss from noise exposure. In C. Harris (Ed.), Handbook of noise control. New York: McGraw-Hill.

Meshkati, N. (1989). An etiological investigation of micro- and macro-ergonomic factors in the Bhopal disaster: Lessons for industries of both industrialized and developing countries. International Journal of Industrial Ergonomics, 4,161-175.

Meshkati, N. (1991). Human factors in large-scale technological systems' accidents: Three Mile Island, Bhopal, Chernobyl. Industrial Crisis Quarterly, 5(2).

Meyer, D., Smith, J., & Wright, C. (1982). Models for the speed and accuracy of aimed movements. Psychological Review, 89,449-482.

Miller, G. (1956). The magical number seven, plus or minus two: Some limits on our capacity for processing information. Psychological Review, 63,81-97.

Mims, F. (1983). Getting started in electronics. Fort Worth, TX: Radio Shack Division of Tandy Corporation.

Mollica, R., Wyshak, G., & Lavelle, J. (1987). The psychosocial impact of war trauma and torture on Southeast Asian refugees. American journal of Psychiatry, 144,1567-1571.

Montgomery, J. (1995). The challenge of literacy. ASHA, May 5.

Morales, R., & Lombard, G. (1986). Feeding aids for the physically disabled geriatric patient. Nursing Homes, March-April, 35-40.

Moriarty, S., & Scheiner, E. (1984). A study of close-set type. Journal of Applied Psychology, 69, 700-702.

Morrison, R., Swope, J., & Halcomb, C. (1986). Movement time and brake pedal placement. Human Factors, 28(2), 241-246.

Moses, R. (Ed.). (1981). Adler's physiology of the eye (7th ed.). St. Louis, MO: V. Mosby.

Moussa-Hamouda, E., & Mourant, R. (1981). Vehicle fingertip reach controls — Human factors recommendations. Ergonomics, 12, 66-70.

Muckler, F., & Seven, S. (1990). National education and training. In H. Booher (Ed.), MANPRINT: An approach to systems integration. New York: Van Nostrand Reinhold.

Mullick, A. (1993). Bathing for older people with disabilities. Technology and Disability, 2(4), 19-29.

Musselwhite, C., & King-DeBaun, P. (1996). Emergent literacy success: Merging technology with whole language for students with disabilities. Birmingham, AL: Southeast Augmentative Communication Publications.

Musselwhite, C., & St. Louis, K. (1988). Communication programming for the severely handicapped: Vocal and non-vocal strategies. (2nd ed.) Boston: College Hill.

Nakaseko, M., Grandjean, E., Hunting, W., & Gierer, R. (1985). Studies on ergonomically designed alphanumeric keyboards. Human Factors, 27(2), 175-187.

Nelson, C. (1991). Practical procedures for children with language disorders. Austin, TX: Pro-Ed.

Nemeth, S., & Blanche, K. (1982). Ergonomic evaluation of two-hand control location. Human Factors, 24(5), 567-571.

Neugarten, B. (1974, September). Age groups in American society and the rise of the young-old. Annuals of the American Academy, 187-198.

Nguyen, T. (1995). Outreach and service delivery to the Southeast Asian population: Reactor statement. In L. Travelli Vitaliti & E. Bourland (Eds.), Project Reaching Out: Proceedings of the Forum on Human Diversity (pp. 50-58). Arlington, VA: RESNA.

Norman, D. (1988). The psychology of everyday things. New York: Basic Books.

Norman, D. (1993). Things that make us smart: defending human attributes in the age of the machine. Perseus Books. Reading, MA.

Norman, D. (2013). The Design of Everyday Things: Revised and Expanded Edition. Kindle Direct. www.amazon.com

Norman, D. (2018). What Went Wrong in Hawaii, Human Error? Nope, Bad Design. Retrieved February 1, 2022. www.fastcompany.com

Noyes, J. (1983). Chord keyboards. Applied Ergonomics, 14, 55-59.

Noyes, J., Haigh, R., & Starr, A. (1989). Automatic speech recognition for disabled people. Applied Ergonomics, 20(4), 293-298.

Nussbaumer, L. (2018). Human factors in the built environment. www.books.goggle.com

Oktay, J. (1985). Maintaining independent living for the impaired elderly; The role of community support groups. Aging, 349, 14-18.

Orr, A., & Piqueras, L. (1991). Aging, visual impairment, and technology. Technology and Disability, 1(1), 47-54.

Ostrom, L. (1993). Creating the ergonomically sound workplace. San Francisco: Jossey-Bass.

Ourand, P. (1995). Curriculum on human diversity. Arlington, VA: RESNA.

Ourand, P. (1997). Personal communication (assistive technology specialist and speech-language pathologist, Morse 2000 World Conference, Minneapolis, MN).

Owsley, C., Sekuler, R., & Siemsen, D. (1983). Contrast sensitivity throughout adulthood. Vision Research, 23, 689-699.

Papenek, V. (1973). Design for the real world. New York: Bantam Books.

Parker, M., & Thorslund, M. (1993). The provision of assistive technology for the elderly of Sweden. Technology and Disability, 2(2), 45-49.

Pennsylvania Assistive Device Center, (c. 1988). Guidelines for assessment and evaluation of students with augmentative and alternative communication needs. Pennsylvania Special Education, Elizabethtown, PA (videotape).

PennTech. (1994). Assistive technology statewide support initiative: Service delivery. Unpublished paper, PennTech: Harrisburg, PA.

Petersen, D. (1984). Human-error reduction and safety management. New York: Aloray.

Petroski, H. (1993). The evolution of useful things. New York: Knopf.

Philippen, D., Ing, D., & Marx, L. (1991). Applicable barrier-free concepts adaptable to modern technology developments. Journal of Rehabilitation Research and Development, 28{1), 206.

Phillips, B. (1991). Technology abandonment: From the consumer point of view. Washington, DC: Request Publication.

Phillips, R. (1979). Why is lower case better? Applied Ergonomics, 10, 211-214.

Pitts, D. (1982). The effects of aging on selected visual functions: Dark adaptation, visual acuity, stereopsis and brightness contrast. In R. Sekuler, Kline, & K. Dismukes (Eds.), Aging in human visual functions. New York: Liss.

Platt, P. (1996). Foreign exchanges. USAir Magazine, February, 50-51, 65-66.

Pomerantz, J. (1981). Perceptual organization in information processing. In R. Kubovy & J. Pomerantz (Eds.), Perceptual organization. Hillsdale, NJ: Erlbaum.

Practical color television for the service industry. (1954). RCA Service Company, Inc.

Prentke-Romich Company. Full line catalog. (1997). Wooster, OH: Author.

Proctor, R. W. & Van Zandt, T. (2008). Human Factors in Simple and Complex Systems. Boca Raton, FL: CRC Press

Proctor, N., Wells, M,, & Seekins, T. (1992). The Range Exchange: A rural assistive technology outreach program by Extension homemakers. American Rehabilitation, 18{1), 14-22.

Public Law 100-407. (1988). Technology-Related Assistance for Individuals with Disabilities Act of 1988. Washington, DC: U.S. Congress.

Public Law 101-336. (1990). Americans with Disabilities Act of 1990. Washington, DC; U.S. Congress.

Public Law 101-476. (1990). Individuals with Disabilities Education Act (IDEA) of 1990. Washington, DC: U.S. Congress.

Radwin, R., Vanderheiden, G., & Lin, M. (1990). A method for evaluating head-controlled computer input devices using Fitts' law. Human Factors, 32, 423-438.

Ramirez, A. (2016). The Alchemy of Us. Boston: MIT Press.

Ramsey, J. (1985). Ergonomic factors in task analysis for consumer product safety. Journal of Occupational Accidents, 7,113-123.

Rao, P., Goldsmith, T., Wilkerson, D., & Hildebrandt, L. (1992). How to keep your customer satisfied: Consumer satisfaction survey. Hearsay. Journal Ohio Speech & Hearing Association, 7(1), 35.

Rathbone, A. (1995). Windows 95 for dummies. Foster City, CA: IDG Books Worldwide.

Reed, P. (1997). Personal communication (director of Wisconsin Assistive Technology Initiative [WATI], Amherst, WI).

Reynolds, R., White, R., Jr., & Hilgendorf, R. (1972). Detection and recognition of color signal lights. Human Factors, 14(3), 227-236.

Richardson, B., Wuillemin, D., & Saunders, F. (1978). Tactile discrimination of competing sounds. Perception and Psychophysics, 24, 546-550.

Ripich, D. (Ed.). (1991). Handbook of geriatric communication disorders. Austin, TX: Pro-Ed.

Rogers, J. (1985). Low technology devices. Generations, 10(1), 59-61.

Rogers, J., & Holm, M. (1991). Task performance of older adults and low assistive technology devices. International Journal of Technology and Aging, 4(2), 93-106.

Romich, B. (1992). The AAC market: Trends and influences. Paper for NIDRR Consensus Validation Conference—Augmentative and Alternative Communication Intervention.

Romich, B. (1994). Knowledge in the world vs. knowledge in the head: The psychology of AAC systems. Communication Outlook, 16(2), 19-21.

Romich, B. (1997). Personal communication (engineer and product developer/manufacturer, Minneapolis, MN).

Romski, M., & Sevcik, R. (1996). Breaking the speech barrier. Baltimore, MD: Paul H. Brookes.

Roper, K. (1997). Learned helplessness. Unpublished research presentation. University of Wisconsin-Eau Claire, Eau Claire, WI.

Ross Thomson, B. (1988). Personal communication (state department of public instruction consultant in special education, Madison, WI).

Rouse, D. (1986). Technology transfer. Generations. Fall, 15-17.

Sanford, J. (2016). Universal design as a human factors approach to return to work interventions for people with a variety of diagnoses. Handbook of Return to Work, pp 403-419.

Salthouse, T. (1982). Adult cognition: An experimental psychology of human aging. New York: Springer-Verlag.

Sanders, M., & McCormick, E. (1993). Human factors in engineering & design. New York: McGraw-Hill.

Sandoz, M. (1961). These were the Sioux. Hastings, SD.

Sarma, J. (1994). Driver configurations for liquid crystal displays. Sunnyvale, CA; OKI Semiconductor.

Schauer, V. (1997). Normal communication development in the elderly as related to conceptual frameworks for aging. Unpublished manuscript. University of Wisconsin-Eau Claire, Eau Claire, WI.

Schewe, C,, & Meredith, G. (1994). Digging deep to delight the older consumer. Symposium on Product Designs and Technologies for the Mature Market, Washington, DC.

Schieber, F., & Kline, D. (1989). Technology for the elderly. SOMA Engineering for the Human Body, 3(1), 15-25.

Schoenfeld, B. (Ed., et al, 2022). ARRL Handbook. Newington, CT: American Radio Relay League.

Schwab, L. (1983). Developing programs in rural areas. In Independent living for physically disabled people. San Francisco: Jossey-Bass.

Seelman, K. (1993). Assistive technology policy: A road to independence for individuals with disabilities. Journal of Social Issues, 49(2), 115-130.

Shadden, B. (Ed.) (1988). Communication behavior and aging: A sourcebook for clinicians. Baltimore: Williams & Wilkins.

Shahnavaz, H. (1982). Lighting conditions and workplace dimensions of VDU-operators. Ergonomics, 25,1165-1173.

Shahnavaz, H., & Hedman, L. (1984). Visual accommodation changes in VDU operators related to environmental lighting and screen quality. Ergonomics, 27,1071-1082.

Shames, G., Wiig, E., & Secord, W. (1994). Human communication disorders: An introduction. New York: Merrill.

Shane, H. (1986). Goals and uses. In S. Blackstone (Ed.), Augmentative communication: An introduction (pp. 29-47). Rockville, MD: American Speech- Language-Hearing Association.

Shane, H., & Roberts, G. (1988). Technology to enhance work opportunities for persons with severe disabilities. In W. Kaiernan (Ed.), Economics, industry, and disability. Baltimore, MD; Paul H. Brookes.

Sherr, E. (1979). Electronic displays. New York: Wiley.

Sherrill, C. (1976). Adapted Physical Education and Recreation. Dubuque; Wm. C. Brown.

Shimon, D. (1991). Coping with hearing loss and hearing aids. San Diego, CA; Singular Publishing Group.

Shorrock, S. (2017). Four Kinds of 'Human Factors': 2. Factors of Humans.

Part two of four-part series. Retrieved December 23, 2021. www.humanisticsystems.com

Shurtleff, D. (1967). Studies in television legibility: Review of literature. Information Display, 4, 40-45.

Shute, S., & Starr, S. (1984). Effects of adjustable furniture on VDT users. Human Factors, 26(2), 157-170.

Siegel, A., & Wolf, J. (1969). Man-machine simulation models: Performance and psychological interactions. New York: Wiley.

Silverman, F. (1995). Communication for the speechless (3rd ed.). Boston: Allyn and Bacon.

Silverstein, B., Fine, L., & Armstrong, T. (1987). Occupational factors and carpal tunnel syndrome. American Journal of Industrial Medicine, 11, 343-358.

Simmonds, G., Galer, M,, & Baines, A. (1981). Ergonomics of electronic displays. Technical Paper Series 810826. Warrendale, PA: Society of Automotive Engineers.

Simpson, C., McCauley, M., Roland, E., Ruth, J., & Williges, B. (1987). Speech controls and displays. In G. Salvendy (Ed.), Handbook of human factors (pp. 1490-1525). New York: Wiley.

Singleton, W. (1982). The body at work: Biological ergonomics. London: Cambridge University Press.

Smith, S. (1981). Exploring compatibility with words and pictures. Human Factors, 23, 305-315.

Smith, S. (1995). Outreach and service provision to persons with disabilities residing in rural and sparsely populated areas: Reactor statement. In L. Travelli Vitaliti & E. Bourland (Eds.), Project Reaching Out: Proceedings of the Forum on Human Diversity (pp. 129-139). Arlington, VA: RESNA.

Smithsonian book of invention. (1978) New York: Norton. Author.

Sotnik, P. (1995). Outreach and service delivery to the Southeast Asian populations of the United States. In L. Travelli Vitaliti & E. Bourland (Eds.), Project Reaching Out: Proceedings of the Forum on Human Diversity (pp. 7-49). Arlington, VA; RESNA.

Spacey, J. (2016). 22 Types of Motivation. Retrieved September 10, 2021. Simplicable. www.simplicable.com

Spacey, J. (2015, updated 2021). 44 Human Factors in Design. Retrieved

February 2, 2022. www.simplicable.com

Spencer, L. (1997). Personal communication (experienced assistive technology user. New Richmond, WI).

Sprott, R., Warner, H., & Williams, T. (1993). The biology of aging. New York: Springer.

Stainback, S., & Stainback, W. (1992). Curriculum considerations in inclusive classrooms. Baltimore, MD: Paul H. Brookes.

Stammerjohn, L., Smith, M., & Cohen, B. (1981). Evaluation of workstation design factors in VDT operations. Human Factors, 23, 401-412.

Stanton, N. A. (2005). Human Factors Methods: A Practical Guide for Engineering and Design. Ashgate Publishing. Retrieved February 1, 2022 via www.goodreads.com

Stern, H. (1979). Rehabilitation and chronic illness in American culture: The cultural psychodynamics of a medical and social problem. Journal of Psychological Anthropology, 1(3), 153-176.

Stein, H. (1995). Rural culture and disability. In L. Travelli Vitaliti &c E. Bourland (Eds.), Project Reaching Out: Proceedings of the Forum on Human Diversity (pp. 199-214). Arlington, VA: RESNA.

Steinfeld, E., & Shea, S. (1993). Enabling home environments: Identifying barriers to independence. Technology and Disability, 2(4), 69-79.

Stevens, J. (1982). Temperature can sharpen tactile acuity. Perception and Psychophysics, 31, 577-580.

Stillman, P. (1994, September-October). The quad squad: A little closer to the front line. A report from the Pine Ridge Reservation. New Mobility.

Stoll, C. (1995). Silicon snake oil: Second thoughts on the information highway. New York: Doubleday.

Stone, R., Staissey, N., & Sonn, U. (1991). Systems for delivery of assistive equipment to elders in Canada, Sweden & the United States. International Journal of Technology and Aging, 4(2), 129-140.

Stuart, S. (1993). AAC issues across the age span. Technology and Disability 2(3), 19-31.

Sudhakai, L., Schoenmarklin, R., Lavender, S., & Marras, W. (1988). The effects of gloves on grip strength and muscle activity. Proceedings of the Thirty-Second Annual Meeting of the Human Factors Society (pp. 647-650). Santa Monica: CA: Human Factors Society.

Talamo, J. (1982). The perception of machinery indicator sounds. Ergonomics, 25, 41-51.

Tenhoor, W., & Shalit, A. (1985). Independence for the elderly: The challenge to technology and business. Aging, 349, 26-30.

Thomason, T. (1994). Native Americans and assistive technology. In H. J. Murphy (Ed.), Technology and persons with disabilities (pp. 379-381). Northridge: California State University.

Timiras, P. (Ed.). (1994). Physiological basis of aging and geriatrics. (2nd ed.). Boca Raton, FL: CRC Press.

Tirado, K. (1997). Morse is the easy part. MORSELS, 3(1), 1-2. Newsletter of the Morse 2000 Worldwide Outreach, University of Wisconsin-Eau Claire.

Torgerson, J. E. (1993). Personal communication. Farmer, welder, machinist, metal artist. Menomonie, WI.

Travelli Vitaliti, L. (Ed.). (1995). Curriculum on human diversity. Arlington, VA: RESNA.

Travelli Vitaliti, L., & Bourland, E. (Eds.). (1995). Proceedings of the Forum on Human Diversity. Project Reaching Out. Arlington, VA: RESNA.

Trease, G. (1985). TIMECHANGES: The evolution of everyday life. New York: Warmick Press.

Tullis, T. (1981). An evaluation of alphanumeric, graphic, and color information displays. Human Factors, 25, 541-550.

Tullis, T. (1983). The formatting of alphanumeric displays: A review and analysis. Human Factors, 25, 657-682.

Tullis, T. (1986). A system for evaluating screen formats. Proceedings of the Thirtieth Annual Meeting of the Human Factors Society (pp. 1216-1220). Santa Monica, CA: Human Factors Society.

Ursic, M. (1984). The impact of safety warnings on perception & memory. *Human Factors, 26,677-6S2.*

United Cerebral Palsy of West Central Wisconsin (1995). Techtots Library. Pamphlet. Eau Claire, WI: Author.

U.S. Department of Health and Human Services, Public Health Service, Indian Health Service (IHS). (c 1997) Trends in Indian health. Washington, DC: U.S. Government Printing Office.

Van Tatenhove, G. (1986). Development of a location, color-coded Etran. In S. Blackstone (Ed.), Augmentative communication: An introduction. Rockville, MD: American Speech-Language-Hearing Association.

Van Tatenhove, G. (1993). Personal talk. Speech-language consultant, Eau Claire, WI).

Vanderheiden, G., & Lloyd, L. (1986). Communication systems and their components. In S. Blackstone (Ed.), Augmentative communication: An introduction (pp. 49-161). Rockville, MD: American Speech-Language-Hearing Association.

Vanderheiden, G., & Yoder, D. (1986). Overview. In S. Blackstone (Ed.), Augmentative communication: An introduction (pp. 1-28). Rockville, MD: American Speech-Language-Hearing Association.

Vash, C. (1981). The psychology of disability. Springer Series on Rehabilitation. New York: Springer.

Von Tetzchner, S., & Martinsen, H. (1992). Symbolic and augmentative communication (trans. Kevin M. J. Quirk). San Diego, CA: Singular Publishing Group. (Translated by Kevin M. J. Quirk.)

Vue, P. (1995). Personal communication (Hmong coordinator. UW-Eau Claire).

Warland, A. (1990). The use and benefits of assistive devices and systems for the hard-of-hearing. Scandinavian Audiology, 29(1), 59-63.

Washburn, S. (1960). Tools and human evolution. Scientific American, 203, 3-15.

Waterworth, J. (1983). Effect of intonation form and pause duration of automatic telephone number announcements on subjective preference and memory performance. Applied Ergonomics, 14(1), 39-42.

Waterworth, J., & Thomas, C. (1985). Why is synthetic speech harder to remember than natural speech? In CHl-85 proceedings. New York: Association for Computing Machinery.

Watson, A., & Collins, R. (1993). Culturally sensitive training for professionals. Journal of Vocational Rehabilitation, 3, 38-45.

Webster, J. (1969). Effects of noise on speech intelligibility. ASHA Report 4. Washington, DC: American Speech and Hearing Association.

Webster's New School and Office Dictionary. (1969). Greenwich, CT: Fawcett Publications.

Weinert, C., & Long, K. (1987). Understanding the health care needs of rural families. Family Relations, 36, 450-455.

Welford, A. (1976). Ergonomics: Where have we been and where are we going? Ergonomics, 19(3), 275-286.

Welford, A. (1981). Signal, noise, performance, and age. Human Factors, 23(1), 97-109.

Wessner, C. (1997). How do the touchscreens on pen computers work? Popular Science, April, 74.

Westbrook, M., Legge, V., & Pennay, M. (1992). Attitudes toward disabilities in a multi-cultural society. Social Science Medicine, 36, 615-623.

Wheatley, A. (2018). Human Factors in Assistive Technology. Retrieved from Alex W. Wheatley online site. January 28, 2022. www.alexwheatley.com

Wickens, C. (1986). The effects of control dynamics on performance. In K. Boff, L. Kaufman, & J. Thomas (Eds.), Handbook of perception and human performance. Vol. 2: Cognitive processes and performance. New York: Wiley.

Winkless, L. (2022). Sticky: The Secret Science of Surfaces. London: Bloomsbury Sigma.

Wilkins, A., & Nimmo-Smith, M. (1987). The clarity and comfort of printed text. Ergonomics, 30, 1705-1720.

Wirkus, M. (1997). Personal communication. Speech-language pathologist and assistive technology specialist with Wisconsin Assistive Technology Initiative [WATI], Amherst, WI.

Wisconsin Council on Developmental Disabilities. (1992). Wisconsin's Motorized Wheelchair Lemon Law. Pamphlet. Madison, WI: Author.

Wisconsin Optometric Association. (1996). The reporter's and editor's guide to eye care providers. Madison, WI: Author.

Wolfgang, L., Kearman, J., & Kleinman, J. (Eds.). (1993). Now you're talking. Newington, CT: American Radio Relay League.

Woodson, W. (1963). Human engineering design standards for spacecraft controls and displays. General Dynamics Aeronautics Report GDS-63-0894-1. Orlando, FL: National Aeronautics and Space Administration.

World Health Organization (WHO). (c 2010). Understanding disability. Author. Retrieved February 1, 2022. www.who.int

World Health Organization. (2021). Disability & health. Retrieved February 3, 2022. www.who.int

Worringham, C., & Beringer, D. (1989). Operator orientation and compatibility in visual-motor task performance. Ergonomics, 32(4), 387-400.

Wray, L., & Torres-Gil, F. (1992). Availability of rehabilitation services for elders: A study of critical policy and financing issues. Generations, 16(1), 31-36.

Wright, T. (1993). African Americans and the public vocational rehabilitation system. Journal of Vocational Rehabilitation, 3, 20-26.

Wright, T. (1995). Look at it from a different angle: A focus on human diversity through Project Reaching Out. In L. Travelli Vitaliti & E. Bourland (Eds.), Project Reaching Out: Proceedings of the Forum on Human Diversity (pp. 1-6). Arlington, VA: RESNA.

Yorkston, K., Honsinger, M., Dowden, P., & Marriner, N. (1989). Vocabulary selection: A case report. Augmentative and Alternative Communication, 5, 101-108.

Zangari, C., Lloyd, L., & Vicker, B. (1994). A historic perspective of augmentative and alternative communication. Augmentative and Alternative Communication, 10(1), 27-59.

Zola, I. (1992). Self, identity and the naming question: Reflections on the language of disability. Social Science Medicine, 36, 167-171.

TERMS AND VOCABULARY

Activation site(s). The place(s) on a device that is(are) touched, tapped or pressed to operate the device.

Adapted play and recreation. Use of technology to allow a child or adult to participate in fun and avocational activities for leisure and learning.

Affordance(s). What the nature, composition, and/or construction of a device indicates about how the device should be used and by whom.

Aided. Using equipment or devices external to the body to assist and augment body skills or movements.
Exosomatic.

Alphanumeric. Displaying letter and number characters on a keyboard, screen or other device in any order. (A B C... *1, 2, 3.. R/Q?X8.... etc.*)

Array. An organized display of symbols, pictures, letters or other information on an AT device; often in a row-column matrix.

Assessment. Collection of data about a consumer, student, client, patient, device, or system.

Assisted listening or seeing. Devices that enhance a user's hearing or vision capabilities.

Assistive technology (AT). In its most basic sense, any tool or device that helps any person to accomplish a task. More formally, AT has been defined as "any item, piece of equipment, or product system, whether acquired commercially off the shelf, modified, or customized, that is used to increase, maintain, or improve the functional capabilities of individuals with disabilities" as described by Public Law 100-407, commonly known as the Tech Act of 1988. This definition was adapted slightly in Public Law 101-476 of 1990, the IDEA and its revisions, by substituting children in uses of individual by the previous act. AT is a highly eclectic field of academic study and clinical practice, enlisting the participation of many professions, and is now constantly influenced by changing and evolving state, national and international laws, rulings, and regulations.

Augmentative and alternative communication (AAC). Support for and /or replacement of natural speaking, writing, typing, and telecommunication capabilities that do not fully meet the communicator's needs. AAC is a subset of AT and is an eclectic field of academic study and clinical practice, combining the expertise of many professions. AAC may include unaided and aided approaches.

Baker's basic ergonomic equation (BBEE). A formula expressing how success or failure with AT may be related to the user's motivation, as well as to physical, linguistic, sensory and cognitive effort of the user, plus time.

Border. A surrounding frame of lines or designs to demarcate a word, icon, or other area on a computer screen or other array.

Cathode ray tube (CRT) screen. A type of visual display in which an electron gun directs a beam of electrons against a phosphor-dot screen that glows from impact of the scanning beam.

Central processing unit (CPU). The main body of the computer, containing the motherboard, drives, and other electronic components essential to storage and processing of information.

Compensation. A type of intervention in which methods, strategies, and / or technologies may be used to take the place of or make up for a limitation or disability of the user. Compensation is often the role of assistive technology.

Composition. The materials or substances of which an item is made.

Constraints. The overt or implicit limitations of how a technology can and cannot be used or operated.

Construction. The maimer in which a device or technology is assembled.

Contrast. How distinct a stimulus is from its auditory, olfactory, tactile, and/or visual background.

Control. A mechanical device used to operate or modify the function of another mechanical or hydraulic device.

Control site(s). The part(s) of the body used to activate switches or controls to operate a device.

Cosmesis. The appearance of an item, and how it relates cosmetically to the user.

Device. Any tool or technology item that can be used to accomplish a purpose.

Diagnosis. Discovering and describing capabilities of a technology user.

Display. Visual portrayal of information on a screen or other medium.

Dvorak keyboard. A computer or typewriter keyboard that is configured with the more frequently used letters and characters grouped nearer the center or "home row" of the keyboard. Use of a Dvorak keyboard can be less exerting while typing because less distance is traveled by fingers, mouth stick, or head stick than with a QWERTY keyboard, where the letters are purposely spread apart.

Dwell. The length of time that a body part, light beam, or other activation source must remain in contact with a switch to operate it.

Dynamic display. The information portrayed visually on a screen, such as pictures or symbols, that can be programmed to change and move without physically altering the screen itself.

Endosomatic. Contained as part of the body, within the body. Unaided.

Environmental control. (EC) Modifying and adjusting important aspects of one's personal environment such as heating, cooling, lighting, radio, television, door opening and closing, water flow, and others.

Environmental control unit (ECU). A device that allows for altering one's environment via remote means, commonly by means of ultrasonic infrared or UHF radio signals sent to receiving units.

Ergonomics. Human factors. The study and practice of how humans interrelate and interact with technology. Human factors tends to be used more in North America. Ergonomics tends to be used more in Europe. The terms are essentially synonymous.

Evaluation. Making value judgments about data collected during assessment.

Exosomatic. Not part of the body; outside the body. Aided.

Feedback. Communication offered by a device to its user; can be auditory, kinesthetic, olfactory, tactile, visual, or combinations of these.

Fitts' Law. A trade-off equation in human factors relating speed of control site movement and area of control site to accuracy of targeting.

Flicker. The regular (or irregular) flashing of a CRT screen, fluorescent lighting, or other sources of illumination due to pulsed electron flow to refresh the glow of the device.

Focus. How sharp or blurred an image appears on a screen.

Forcing function. "Fail-safe" measures built into technology to ensure safe use and to prevent or limit accidents or damage to device and user.

Glare. Direct light from sun or artificial sources or reflected light from smooth or shiny surfaces such as screens that can impair the user's ability to clearly see information displays and to interact with a device.

High or Hi tech. Typically complex, expensive electronic and/or mechanical/hydraulic equipment; often translucent or opaque in actual use. Commonly contains digital electronics, microprocessor devices, and power supplies.

Intervention. Actions taken on behalf of a student, client, or patient to help him/her; can be of a variety of types: remediation, compensation, management, and others.

Keyboard. The input device showing letters, numbers, and function commands for communicating with a computer or typewriter; the array of controls or switches for operating a piano, organ, synthesizer, or similar musical instrument.

Knowledge in the head vs. in the world. Operational understanding and knowledge of what to do with a device can reside with users from their prior learning and experience, and/or it can be evident on the keyboard or other input device used to operate technology (e.g., Norman, 1988, 1993 and 1998; Romich, 1994).

Label. A surrounding field of contrasting color, tint, or texture used to highlight words, icons, symbols or other information on a screen or other display.

Latching. A feature built into a switch or control that allows it to stay in a constant state of "on" or "off" until activated further.

Learned helplessness. Inactivity and passivity in children or adults, perhaps living with disabilities or other challenges; it can become reinforced and entrenched due to repeated failures or lack of stimulation in interacting with people and devices.

Lifespan. The total changes, challenges, and expectations a human encounters from birth to death; may be around 80 to 100 + years for persons in many countries.

Light-emitting diode (LED) screen. A visual display in which an array of electronic components (semiconducting diodes) are configured in combination, serving as pixels to create onscreen images. LED. OLED and AMOLED screens are made to glow by passing electric current through them.

Liquid crystal display (LCD) screen. A visual display on calculators, smart phones, tablets, laptop computers, and other devices in which images are formed via electronic modification of light transmission characteristics through liquid crystal compounds embedded in a backlit screen.

Literacy. Capability to use speech, hearing, writing, and reading as tools for thinking.

Lite or light tech. See Low tech.

Load. Amount of physical, cognitive, sensory and/or linguistic effort or exertion in a task; can also include time duration of a task.

Low tech. Technology that involves common, simple, readily available, and usually inexpensive components, typically transparent or translucent in use, and may even involve unaided, endosomatic strategies.

Management. Intervening in ways that help individuals, their family and care providers learn to live day-to-day with a condition or circumstance, rather than trying to remediate it or compensate for it.

Mappings. The pattern and flow of movements to operate or use a device. Mobility Moving within and among environments.

Mounting. How a switch, control, or other device is secured or attached so that a user may gain access to it to operate a technology.

No tech Strategies. Methods, strategies, and/or simple, low-cost devices that allow a user to accomplish a purpose, usually transparent or translucent in device operation. Unaided.

Opacity/opaque. Device operation characteristics that require considerable training, prior knowledge, or experience with a required system of rules and skills. Can be perceived as user-unfriendly, or hard to use.

Ordered sets. Information displayed on a device grouped in logical, predictable ways that help the user to remember and follow operational commands or sequences.

PDA. Personal data assistant. Handheld or pocket digital device to store and manage data, dates, information.

Palliative intervention. Intervention that is focused on easing symptoms and on making a condition easier to live with for a patient or client; neither remediation nor compensation for the condition is intended.

Positioning. How device users are seated or otherwise oriented so that they can see, manipulate, and otherwise interact with toys, tools, and other technologies. Positioning for general consumers and AT users may vary considerably across tasks, settings, and times of day. Positioning also pertains to the switches, controls, and devices a user may be interacting with.

Powered mobility. Wheelchairs, scooters, and other devices that rely on electrical or other power sources to propel them under the user's control.

Prosthesis. A device fashioned to replace or supplement the function of a missing or weakened limb or other body parts. Selecting, fitting, and training persons in the use of their prostheses is the profession of prosthetics.

Proxemics. Considerations regarding distance between user and device, or user and device and other people.

QWERTY The most common keyboard array used with computers and typewriters. Q-W-E-R-T-Y are the first six letters of the upper left-hand row of letter keys.

Random sets. When information displayed on a device is not grouped in an organized, logical manner; may have an adverse impact on mapping for device use.

Rehabilitation robotics. Use of three-dimensional devices that move according to user commands to aid persons in meeting personal care, environmental control, recreational, vocational, and other needs.

Resolution. Regarding switches and controls: how distinct, obvious, and clear an activation site and the information displayed on it are—usually related to the surface area and size of an activation site. Regarding screens: how distinct, obvious, and clear the information displayed on the screen and related image details are, particularly as images are reduced or enlarged in size.

Screen. Visual display device that is self-illuminated or that reflects light to convey information via text or other images; common types in AT are LCD, LED, CRT, Plasma, Flexible, and other, newer devices.

Selection technique. Way information is communicated to a device. Varieties can include direct, indirect, scanning, and encoding, and combinations thereof.

Selectivity. How likely it is that a given activation site may be touched or otherwise selected accurately by using a given control site without mistakenly activating other neighboring activation sites.

Sensitivity. The amount of muscle force needed to activate a switch or control; can also equate with dwell time in optical pointing.

Sensory defensiveness. A reluctance or inability of some technology users to tolerate contact of their skin or other body parts with switch, control, or other surfaces; also referred to as tactile defensiveness. May relate to texture, temperature, or other aspects of surfaces that contact a user's body.

Smart phone. A multifunction telephone, camara, calculator, texting device, and more, allowing internet access among other features in a compact, handheld form, typically employing a flat, touch-sensitive screen.

Smart speaker. A hands-free device that searches the internet and allows telecommunications via speech and voice access.

Speech recognition technology (SRT). Devices are activated and controlled primarily by the user's speech and/or voice.

Static display. A fixed visual or tactile display of letters, numbers, pictures, icons, or symbols; it cannot be changed without physically removing and replacing items on the display.

Switch. An electronic or electromechanical device used to interrupt, control, adjust, or modify electron flow in an electrical circuit.

System. From a human factors perspective, a person using a device to accomplish a purpose. All three components are necessary for a true system to exist.

Tablet. A pad of paper sheets fastened together for note taking, useful as low-tech communication tool. Also, a portable, flat screen device functioning as a combination of a smart phone with a notebook computer.

Tactile defensiveness. See Sensory defensiveness.

Technological literacy. Experience with and knowledge of technology applications and operation that accumulates for a user, and that can allow him or her to generalize certain skills from one device to another.

Technology. Strategies and methods that may or may not be used with tools, devices and equipment which allow a user to accomplish an intent or task.

Tool. An item, object or device used by a human or animal to accomplish a purpose that could not be accomplished as well, as safely, or at all by using just a part of the body.

Translucent. When the method of use of a tool or other device can be inferred and understood once a basic rationale of operating skills or instructions are provided for a new user. These may not be obvious at first, requiring some explanation or training to allow the consumer to make use of the device.

Transparent. When the method of use of a tool or other device is obvious, even to a new user. Transparent technologies are often referred to as "user-friendly," visible, or open.

Unaided. Using the body only; not requiring external devices or equipment. Endosomatic.

Variability. The capability of a switch or control to change the amount of electrical current flow or mechanical resistance over a range of possibilities, rather than just in binary on/off states.

Video display terminal (VDT). CRT, LCD, LED, OLED, AMOLED or other electronic screens that convey information, often in fixed locations as on a desktop computer, airport data display, or in other applications.

Voice input or recognition device. Human speech or other related sounds active and control a technology.

Voice output communication aid (VOCA). A portable electronic device that incorporates synthesized or digitized speech output.

YOUR HUMAN FACTORS TOOLKIT
A PRACTICAL CHECKLIST YOU CAN USE RIGHT NOW

For each device, tool, or technology, apply these Human Factors & Ergonomics Considerations:

OVERALL (From Chapters 1-5)

Transparency, translucency, opacity

Use of ordered sets vs random sets of information to operate

Cosmesis

Mappings

Affordances

Learned or taught helplessness of user

Feedback in multiple modes for multiple senses

Knowledge in the world vs Knowledge in the head to operate

Constraints, including physical, semantic, cultural, and logical

Forcing functions for correct operation and safety

Mistake, error, and misuse prevention

Baker's Basic Ergonomic Equation (BBEE):

```
MOTIVATION TO USE AND SUCCEED
WITH A TECHNOLOGY            (M) = USER SUCCESS...or not (S)
PHYSICAL + COGNITIVE + SENSORY
+ LINGUISTIC + TIME LOADS    (L)
```

SWITCHES AND CONTROLS *(From Chapter 6)*

Sensitivity

Resolution

Contrast

Feedback

Latching and variability

Positioning and mounting

Composition and construction

Tactile and sensory defensiveness of user

Fitts' Law, match of activation sites & control sites:

$MT = a + b \log 2(2D/W)$

SCREENS *(From Chapter 7)*

Size and bulk

Power supply

Proxemics for seating and positioning re screen and controls

Display of visual Information re:

    a. Contrast and brightness, resolution, flicker, and focus

    b. Screen glare and reflectivity

    c. Font sizes, types, contrast, labels, and banners

    d. Feedback for entry errors, misspellings, other device functions

    e. Static or dynamic displays of text, graphics, symbols, icons

RELATED IDEAS FOR TECH LEVELS *(From Chapter 8)*

Can/should your solutions include one for more of these levels of technology? How? Why?

No tech:

Low tech:

High tech:

Combinations and integrations of the above?

How will you configure and apply one, two or all of these approaches?

VARIABLES FOR SUCCESS AND FAILURE *(From Chapter 9)*

Human factors regarding the people surrounding the user:

Human factors regarding the user:

Human factors surrounding the device, tool, or technology itself:

How will each of these factors impact success or failure? Why?

What will you do about each of those impacting factors?

Permission to copy this Tool Kit list is granted by the author. Please show the credit notice below.

Adapted from *ADD HUMAN -- human factors and ergonomics for all of us.*

Copyright 2022 Thomas Wayne King. All Rights Reserved. ISBN 9798496622639

I hear and I forget. I see and I remember. I do and I understand. - Confucius

ABOUT THE AUTHOR

About the author: Thomas Wayne King, Ed.D., is professor emeritus of communication sciences and disorders, University of Wisconsin-Eau Claire, and a retired clinical speech-language pathologist (CCC-SLP, ret). Elected in 1991 as the first president of the Wisconsin Society for Augmentative and Alternative Communication (WISAAC), King is author of many professional & academic books and articles. His next textbook for general audiences is planned for 2022-23 release: *Writing with Lightning -wonder & magic of telegraph codes*. Thomas writes from a lifetime of primary research, with family and professional experience in technologies for all persons, including learners and adopters living with exceptional abilities and challenges. Tom is also a prolific author of other fun books, articles, essays, stories & songs. He holds the highest-class FCC amateur radio license: Amateur Extra. His call sign is WF9I. Tom and Debra, N9GLG, raise registered Icelandic wool sheep at their Sunny Cove Farm overlooking Lake Superior. The Kings run, hike, ski, and skate…and celebrate life daily as they haul feed & bales throughout Northland seasons. They have two grown scientist sons, one a physicist and one a physician, and two amazing, wonderful grandchildren.

SOME OTHER WORKS BY THOMAS WAYNE KING

My works also include many diverse scientific, academic, clinical, historical & popular articles and chapters, plus more than 150 songs, raps and rhymes written and composed across eight decades, most registered with ASCAP.

Textbooks, Books and Chapters:

Assistive Technology: Essential Human Factors. Boston: Simon & Schuster/Allyn & Bacon/Pearson Education, Inc. 1998. ISBN 0-205-27326-2

Modern Morse Code in Rehabilitation and Education: new applications in assistive technology. Boston: Simon and Schuster/ Allyn & Bacon/Pearson Education, Inc. 2000. ISBN 0-205-28751-4

"Pediatric Assistive Technology" Chapter 36 in textbook *Pediatric Life Care Planning and Case Management*, 2nd edition. S. Riddick-Grisham and L. Deming, editors. New York: CRC Press/Taylor Francis Group. 2009. ISBN 978-1-4398-0358-5 (with coauthors A. King & D. King)

"The Technologically-Dependent Child" Chapter 40 in textbook. *Pediatric Life Care Planning and Case Management*, First edition. S. Riddick-Grisham and L. Deming, editors. New York: CRC Press. 2004. ISBN 0-8493-1726-6

Tales from the Red Pump: 130 years of Northland family adventures. Volume 1. Raleigh, NC: Lulu Publishing/ Sunny Cove Publishing. 2009-2016. ISBN 978-0-578-04214-5

Sailor of the Sun -volume 2 of *Tales from the Red Pump* series. Raleigh, NC: Lulu Publishing/ Sunny Cove Publishing. 2010-2017. ISBN 978-1-365-47774-4

Neighborhood of Bears: essays, tales, notes, stories, history. Volume 3 of *Tales from the Red Pump* series. Raleigh, NC: Lulu Publishing/ S9unny Cove Publishing. 2016-2018. ISBN 978-1-365-99556-9

Magic Snow Socks ...tales of practical transcendence. Volume 4 of *Tales from the Red Pump* series. Raleigh, NC: Lulu Publishing/ Sunny Cove Publishing. Solon Springs, WI USA. 2019-2020. ISBN 978-1-67802-766-7

Trash Picker – danger on a tropical beach. Book 1. ISBN 978-1-67802-

779-7 and ***Trash Picker –dangers North. Book 2.*** ISBN 978-1-67803-016-2. Fictional novella series. Amazon Kindle Direct/Sunny Cove Publishing. Solon Springs, WI USA. ***Trash Picker – return to the beach. Book 3.*** In progress for 2022-23 release.

Red Pump Chronicles --130 years of Northland stories, articles, poems, lyrics, raps collected from my ***Tales from the Red Pump*** series, volumes 1, 2, 3 & 4. Raleigh, NC: Lulu Publishing/Sunny Cove Publishing, Solon Springs, WI USA. 2021. ISBN 978-1-67803-012-4

Noon at the Nerd Table -- tales & essays for the curious, passionate & intense. Amazon Kindle Direct/Sunny Cove Publishing. Solon Springs, WI USA. 2021. ISBN 9798517944481

Homestead of the Red Pump. Collection plus my newest Northland works in progress. Amazon Kindle Direct/Sunny Cove Publishing. Solon Springs, WI USA. Planned for 2024 release. ISBN 978-1-67803-939-4

Oh My Gosh... Have We Got Squash! ... tales and charms from gardens and farms. Thomas Wayne King & Debra Raye King, coauthors. Amazon Kindle Direct/Sunny Cove Publishing. Solon Springs, WI USA. 2021. ISBN 9798520644583

ADD HUMAN -- human factors & ergonomics for all of us. Revised enhancement of *Assistive Technology: Essential Human Factors* (1998). Amazon Kindle Direct/Sunny Cove Publishing. Solon Springs, WI USA. 2022. ISBN 978-1-67803-347-79

WRITING with LIGHTNING - wonder & magic of telegraph codes. Revised enhancement of *Modern Morse Code in Rehabilitation and Education: new applications in assistive technology (2000)*. Amazon Kindle Direct/Sunny Cove Publishing. Solon Springs, WI USA. 2022. ISBN 9798496668224

My LP/EP Music & Spoken-word Albums created & written (Reg ASCAP), performed, produced & recorded as of 1/28//2022:

They Run! LP 2014 ***Winter Magic*** LP 2014 ***Sunny Cove Words*** LP 2014 ***Wolf Moon*** LP 2010 ***Good Hay*** LP *2010* ***Wild & Woolly*** LP 2010 ***Tales from the Red Pump volume 1 Audio Book*** 2010 ***Aardvarks & Angels*** 1998 and ***Starlight Dolphin*** 2000 LPs by T. King co-created w/ M. Jahr. ***Lightning Girl Collection*** EP *and* LP for 2023-24 release ***And More Instrumentals, Songs and Raps in Progress…***

MY TRIBUTE TO MALCOLM "MAL" HANCOCK

American comics artist and cartoonist. Mal often spoofed human factors and ergonomics.

May 20, 1936 – February 16, 1998

Mr. Hancock has been one of my favorite cartoonists for decades. I am pleased to honor Mal's memory and life story with his insightful cartoon of the classic human factors & ergonomics dilemma:
Which switch does what? And does it matter?
(Yes, it does! - twk)

Most noted for his work in *National Review*, Malcom "Mal" Hancock's cartoons appeared regularly in *The Saturday Evening Post*, *Playboy*, and the *Washington Post*, among many other publications. Mal also created and drew several original comic strips, each with large, lasting followings. As a teenager, he became a user of assistive technologies after a serious fall, attributing his success as an artist to the focus and determination he gained from his disability.

MY TRIBUTE TO MALCOLM "MAL" HANCOCK

"Mr. Hancock was paralyzed from his waist down and used a wheelchair after a fall he suffered as a teen. He was vacationing with his family in Wisconsin. He and his dad decided to play a round of golf. Mal was looking for a lost ball and did not realize he was so close to the unmarked edge of a limestone cliff. He slid about 15 feet and then free fell the additional 35 feet. He credits his handicap for realizing his potential. Mal said, "You have to play with the hand you are dealt. Everyone has a handicap--with some it is physical; with others, it is psychological, mental or whatever. Sometimes a handicap is an advantage because it focuses your attention on your talent. I was lucky because I had a particular talent that could be focused--and it wasn't tied to my legs." Mal died at age 56 of cancer."

Adapted from Wikipedia citation for Malcom "Mal" Hancock, American artist and cartoonist. Retrieved January 23, 2022.

www.ingramcontent.com/pod-product-compliance
Lightning Source LLC
Chambersburg PA
CBHW071444220526
45472CB00003B/651